ENTERPRISE-FOCUSED MANAGEMENT

Changing the face of project management

The application of the Theory of Constraints
to the enterprise management of companies

DR TED HUTCHIN

 Thomas Telford

Published by Thomas Telford Publishing, Thomas Telford Ltd, 1 Heron Quay,
London E14 4JD.
URL: http://www.thomastelford.com

Distributors for Thomas Telford books are
USA: ASCE Press, 1801 Alexander Bell Drive, Reston, VA 20191-4400, USA
Japan: Maruzen Co. Ltd, Book Department, 3–10 Nihonbashi 2-chome, Chuo-ku,
Tokyo 103
Australia: DA Books and Journals, 648 Whitehorse Road, Mitcham 3132, Victoria

A catalogue record for this book is available from the British Library

ISBN: 0 7277 2979 9

© Ted Hutchin and Thomas Telford Limited 2001.

Typeset by Academic + Technical, Bristol
Printed and bound in Great Britain by MPG Books, Bodmin

Acknowledgements

The author wishes to acknowledge the following people without whom the book would never have seen the light of day. There are those from the construction industry who have been part of the data collection process, and those who have offered advice and observations about what has been written: Martin Print, Neil Moore, Phil Bayliss, Lawrence McKidd, John Cope and Neil Butterill. There are those from the TOC community who have been involved in similar implementations in other parts of the world, and have also given advice and offered suggestions about how to improve the message: Alan Leader, Kathy Austin, Barry Urban, Bert Husken, Shaun Doherty, Danny Walsh, Tali Mastboim, Dennis Marshall, Dick Peschke and, of course, Dr Eli Goldratt. There are those from the wider world of project management who have challenged me and given me data with which to understand the world they live in and try to make money from: Larry Leach, Tony Rizzo, Tom Gronek and the many people from the various companies I have worked with over the past ten years in the implementation of TOC solutions.

No book ever comes to fruition without the help and encouragement of the publishers, I would like to take this opportunity to thank Jeremy Brinton and all the team at Thomas Telford for giving me the ability to speak to a wider audience.

Finally, there are those close to me whose support and guidance were instrumental in keeping me to time and content, particularly my wife Audrey and family, and also my many friends and colleagues around the world, without whom the book would still be an idea, like the dawn waiting for the sun to rise.

Contents

Introduction

This book came about as a result of research I was doing at Cranfield University and as part of my work at I. & J. Munn Ltd. Over the past five years I have been involved in implementing the Theory of Constraints (TOC) to a number of companies throughout Europe and the USA. The major application has been in the field of project management using an approach entitled *critical chain*. I was first introduced to the world of projects in 1971 while serving with the REME in the British Army. As a radar engineer I was part of a team involved in the implementation of a radar system, which was associated with the NATO missile range in the Outer Hebrides of Scotland. Although my association with that project ended back in the 1970s, when I came back to project management with the TOC approach during the 1990s one very interesting dimension stood out, little had changed in the way projects were managed. Although the technology of laying out a network of activities had moved from peg boards to computer screens the use of PERT and/or Gantt charts was much the same. Critical path was still the primary focus of most project managers, and they still seemed to have problems keeping control over the whole of the project. These days, in many companies, there are more projects at any one time, and often people are spread across two or three projects at the same time, and expected to deliver against ever shortening lead times. However the problems of management seem to be the same, only more complex!

The critical chain approach was first described by Dr Eli Goldratt. He and others developed the critical chain concept within the Avraham Y. Goldratt Institute during the early part of the 1990s. I was the consultant for, and facilitated some of the earliest implementations within the UK and Europe, and in particular within the construction industry. More will appear about these as part of the case studies, which form the primary data for this book. The approach has continued to develop at a rapid pace and will no doubt continue to develop over the next few years. The basic principles however should remain the foundation of that development for many years to come. Indeed I would argue that when Dr Goldratt wrote his book entitled *Critical chain* two elements were not given the right level of importance. The first was the impact of lacking a focusing tool within multi-project environments, and the second was the need to have properly constructed networks before applying the critical chain scheduling itself. More about these two dimensions, and the whole of the critical chain approach later in the book.

The book itself follows a path first developed by Dr Eli Goldratt and called *the six layers of resistance to change*. I have developed this approach based on the work I completed at Cranfield and call it the *five steps to successful change*. Although this will be developed in more detail later when describing the implementation of critical chain, the book itself uses the same model, hence a few words of explanation here.

The first step concerns gaining consensus on the problem, which refers to the need for all concerned to agree that the problem is what the analysis suggests. Without consensus on the problem there can be no foundation for solution development. The potential for a chaotic result is high if such consensus is not forthcoming. The concept described by David Ruelle called *sensitive dependence on initial condition* is most relevant here. The solution depends on the level of clarity given to the analysis of the problem and the way in which it is stated. There is also a core element related to the role of clear communication; the importance for everyone to know what is going on. The rigour of the TOC approach to problem solving through the application of the *thinking processes* (TP) that lie at the heart of the TOC offers the only real methodology for reaching such consensus.

The second step is concerned with gaining consensus on the direction of the solution. This refers to the need to understand the broad direction of the solution before the full development of the solution is complete. It is a simple question about whether the direction is right or not; does the solution in this brief form overcome the problems to the agreement of all concerned. Once more there is a high premium on the need for consensus through clear communication.

The third step is gaining consensus on the benefits of the solution, which refers to the full development of the solution with the benefits linked to the changes that will have to take place as part of the implementation. This analysis, again using the full rigour of the TOC/TP, allows for a real debate about how the expected benefit will be achieved and the acquisition once more of real consensus.

Successfully overcoming reservations is a necessary condition if the solution is to be properly implemented. Based on having achieved the consensus of the previous three stages, the type of reservation falls into two distinct categories. The first is the recognition by an individual that if the solution is put into practice it will combine with something that already exists to lead to a more damaging state. This new, and more negative, dimension will outweigh thus the intended solution, and the benefits that were expected. This is called *negative branch reservation* (NBR) and the application of this tool within the TOC/TP is crucial to first understanding the nature of the reservation, and then overcoming it. The second type of reservation is the identification by the individual that there is a barrier; an obstacle to implementing the necessary changes defined in the full solution analysis. These obstacles need to be captured and then each one in turn overcome. The TOC/TP tool for this activity is known as a *prerequisite tree* (PRT). Using both of these tools, the NBR and the PRT, it is possible to successfully overcome the reservations of all the team members. The final tool from the TOC set is the *transition tree*, which outlines each action in the sequence in which they must take place, with the expected outcome made clear. All that is left to do now is to actually implement the solution, i.e. make it happen.

However, even though making it happen is the next obvious step, the research I carried out at Cranfield showed that even when all of the preceding stages have been successfully accomplished nothing happens. The research set out to investigate this problem and discovered the existence of what I call *paradigm lock*. This is where the individual is caught in the current, dominant paradigm about a specific issue and feels unable to change, even when the change is agreed and of benefit. This aspect of change management will be discussed in greater detail later.

There are some basic assumptions about the people who will read this book, and who may even implement some of the ideas and concepts it contains. The first is the level of care they have about their organisation, their team, and the people around them. There is a basic assumption among the people around the world who work in the TOC community, namely, that the people we work with in company care about their people, their customers, their suppliers and so on. Those who care see these people as important contributors to the success of the enterprise and seek to generate win–win relationships with them at all times. This is a fundamental assumption, which applies to all the work I and my team carry out in the various assignments we are involved in around the world.

The second is the intuition they have about their organisation and the environment in which it operates. The TOC/TP is essentially a rigorous thinking process for problem solving and knowledge capture. Although the process is generic in nature many of the problems it addresses are specific to the environment in which they exist. The TOC practitioners cannot always have the same level of intuition that those inside the organisation can have. Therefore the people within the organisation must contribute this aspect in any TOC analysis. Of course, if the TOC practitioner has a level of intuition then the ability to help the customer can be substantially enhanced.

The third assumption is mine. I am firmly of the opinion that the TOC/TP tools described within this book can deliver substantial improvement to the bottom line of the company, and help individuals achieve their own personal goals. When placed in combination with the care and intuition of the people within the organisation there is no problem that cannot be addressed, no market that cannot be attacked, and no decision that cannot be taken in a win–win manner.

So what of the subject matter itself – project management based on the TOC. The years from 1995–2000 witnessed a dramatic rise in the number of companies implementing, and gaining benefit from, the Theory of Constraints approach to project management known as critical chain. The primary sectors of the economy taking advantage of this approach were construction, communications, microprocessor manufacture, aircraft maintenance and healthcare. The geographical spread went from North and South America, South Africa, Pacific Rim, Europe and the UK. The companies ranged from the very big to the very small. The projects ranged from the very small (<£1 million) to the very big (>£200 million). Many of these companies gained substantial bottom-line improvement as a result of their actions. Some gained strategic leverage in their chosen market. But critical chain is not the whole story. This book is an attempt to describe what happened in a number of specific implementations, i.e. the case studies in chapter 4. It is also an attempt to place the critical chain approach into the wider, enterprise dimension of the companies involved.

The intention is to demonstrate first the challenging nature of critical chain and the Theory of Constraints from which it was developed. Second, to show that without an enterprise wide focus, the real pay-off remains just out of reach. Third, that managing the change, which critical chain requires as a necessary condition for success, cannot be achieved without a model of change that recognises the real power of the ruling paradigms to act as a barrier to change itself. Finally, to give the reader the necessary understanding of the ways in which, what I call the *laws of constraints* operate and cannot be ignored, cannot be left unaddressed, if the goal of the company is to be achieved. For the purposes of this book the goal which applies to for-profit companies is defined as the ability to make money now and in the future coupled to two necessary conditions. The first is that the market, from supply base to client, is satisfied now and in the future and the second is that the team within the organisation, however defined, is equally satisfied both now and in the future.

CHAPTER ONE

The current situation within project-based companies

Introduction

This book is about business, about those companies who are trying to make money through the successful completion of projects. This is not an easy place to make money however. The numerous business issues that have to be dealt with on a day-to-day basis seem to cause untold problems and headaches for all levels within these, indeed all, types of companies. Issues such as shareholder value and how to increase it are forcing managers to focus on the financial measurements, and in particular cost, as never before. Other issues such as improving market-share, profitability and productivity are all part of the daily discussion about the organisational welfare and how to improve it. Questions about the effectiveness of the sales force with respect to the competition, questions about the productivity of the people, questions about the size and nature of the cost base and so on, are being asked almost daily in the corridors of power. Answers to other questions such as the ability to stay at the leading edge of technology in terms of both products and the processes to manufacture them are fast becoming necessary conditions for survival and then growth.

These questions and many more are often part of the problems facing managers at all levels, from boardroom to shopfloor. What is equally apparent is that the focus seems to shift from one aspect to another depending on the pressures of the day. We have found in our work around Europe and the USA that many managers are spending a great deal of time bouncing from one issue to another, and trying to deal with them as if they were independent. Even if their intuition tells them that many of these issues are interrelated, the systems within the organisations often do not recognise this fact.

Thus these issues, these questions, are the staple diet of the commentators, the investors, the shareholders and the managers of organisations around the world. Many of the companies trying to deal with these issues, resolving the conflicts they create, are companies involved in projects, and therefore project management. The projects may be related to new product development, civil engineering, a technological break-through and many more, but they all share the same basic requirement, the same necessary condition for success, the delivery of a project, or group of projects on time, in full and to budget. These elements are the necessary conditions for a successful project. Failing to meet these necessary conditions violates the necessary conditions of the organisation in terms of the market, e.g. quality, and in terms of the people within the organisation. This is the way they make their money. This is a key dimension of their goal. Without such success they will suffer the inevitable consequences of non-performance. This book is about such companies, such issues and such conflicts. It is a book that uses the critical chain approach developed by Dr Goldratt to address these problems, these issues. It is a book that argues strongly for a proper alignment of functions, resources and above all measurements, for the organisation to be successful. In other words this book argues for an enterprise-focused approach to the management of projects and project-based companies. However, before the solution can be really discussed and understood it is perhaps necessary to consider the current environment of projects and project management.

The current environment of project-based companies

Ensuring the ability to meet the demand of the market in terms of due date performance and ever shorter lead times, coupled with the never ending pressure on price, means that many companies are trying to focus on how to achieve such levels of performance without risking the company. This is usually linked to the need for very lean environments, as argued by commentators and researchers such as Pascale (1991), Schonberger (1982 and 1986), Womack, Jones and Roos (1990), and Hayes, Wheelwright and Clark (1988). Within the manufacturing sector, the emphasis today is on the ability to bring to market, on time, products that meet specifications and can be made effectively within the production facility without significant problems. Within civil engineering, the emphasis is on the ability to complete the construction project within the budget, without changing the specification, and in terms of time, on or before the due date agreed upon with the client and written into the contract. This emphasis is also closely allied to the goal of the company and the environment in which it exists, which in turn determines its corporate strategy. To be successful in this environment requires the ability of management to see all of the functions of the organisation as part of the whole company and not as a group of single entities, independent and operating almost without reference to the rest of the organisation.

In writing about what they called the new competitive challenge for manufacturing, Hayes and Wheelwright (1984) wrote

> *Studies of manufacturing firms in a variety of countries have per-suaded us that the economic problems facing US companies in the 1980s – and particularly the productivity problem – have been due less to foreign pressure and governmental pressure than to some critical weaknesses in the way that US managers have guided their companies. These weaknesses have called into question some of the basic assumptions and practices that govern the way top US manufacturing companies have reacted to their strategic challenges.*

There is no difference between those companies in sectors such as construction, chemicals and pharmaceuticals and those who deliver projects rather than manufactured products. Although it is always easy to blame others, especially outside the company, the real responsibility may lie within. Hayes and Wheelwright (1984) present a powerful argument for dealing with this problem. They recognise, as does Goldratt (1990), the problems associated with focusing on just the short-term financial measures. They argue for the development of a strategic and coherent management philosophy linked to the overall business objectives.

Ishikawa (1990) considers that

> *Even when good improvement proposals are made, they often cannot be executed satisfactorily. Everyone rushes around shouting that they are eliminating defectives and increasing production, but in the end nothing is improved. This is because they are confusing control with improvement. If we want to make improvements we must first have total control. Only when control is sufficiently well implemented do significant improvements become possible.*

Although referring to quality improvements the lesson is straight-forward. Only when the enterprise is under control and the direction is clear can real improvements be considered. Therefore the ability to control is a key feature of any enterprise trying to achieve a goal, and thus improve the performance towards that achievement. Once the organi-sation is under control, then the two words that should dominate senior management thinking and action are *focus* and *leverage*. Where to focus, indeed how to focus, and what is really meant by leverage will be covered later.

Organisations exist for a purpose. Their owners set up the company in order to provide for a perceived need within the market place. The owners determine the goals of the organisation whether it is making money, serving the customer or any other they deemed. Buchanan and

Huczynski (1985) suggest that organisations 'exist where individuals acting alone cannot achieve goals that are considered worthwhile pursuing'. They go on to confirm that

> *Organisations do not have goals. Only people have goals... Senior Managers may decide on objectives and attempt to get others to agree with them by calling them 'organisational goals'; but they are still the goals of the people who determined them in the first place.*

Porter (1980) considers that 'the essence of formulating competitive strategy is relating a company to its environment'. He then reflects that the goal of the competitive strategy for 'a business unit in an industry is to find a position in the industry where the company can best defend itself against these competitive forces or can influence them in its favor'. Porter suggests that the primary forces driving industry competition are

- suppliers
- potential entrants to the market
- buyers
- rivalry between existing firms
- substitutes, either in terms of products or services.

This is shown in Fig. 1.1.

Given these forces it is essential to develop a clear strategy to combat them. He further argues that the three primary avenues that are successful include the following – overall cost leadership; differentiation, by which

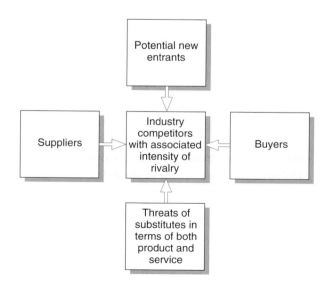

Fig. 1.1. The Porter model

he means the ability to segment the market to develop a unique position; and focus which involves targeting either a specific market or group of people.

Goldratt and Cox (1984) argue that the overall goal of a commercial organisation is the ability to make more money now and in the future, through sales. Compared to Porter this suggests a different focus. Porter argues that the primary focus is cost control. While not suggesting that such approaches are not important, Goldratt argues that just concentrating on cost control is not enough, the focus must be on making money not saving it, and that means sales. If the objective is about sales and therefore about the ability to meet market demand with both current and new projects, then time becomes an important factor. This emphasis on reducing the project cycle time is not without reason. If, in any market, competitors are able to deliver earlier, then any company is immediately at a disadvantage. Slow to market means lost sales, and that in turn can lead to closure. Therefore any technique that can reduce the overall project lead-time is worthy of closer examination. What is already clear is the fact that the existing techniques such as CPM/PERT have been found wanting. Also that most of the current software offerings to the market only enshrine the project management model in use today without challenging any of the basic assumptions it contains.

Tichy (1983) suggests that

> *The argument is made that an effective organisation is one in which there is good strategic alignment, that is the organisational components are aligned with each other, and the political, cultural and technical systems are in good alignment with each other.*

Thus developing a successful strategy involves a clear understanding of the goal of the organisation as determined by the people who own the organisation, and the ability to bring together the various elements of the organisation in a coherent structure. Coherence demands that every part of the organisation is seen as part of a chain rather than as a series of independent links. The strength of the organisation is therefore judged by the strength of the weakest link and it is here that the overall strategy of the organisation should focus.

Any organisation, operating within either the commercial or non-commercial environment, must demonstrate its ability to achieve the goal set for it by the owners. The responsibility for this achievement rests within positions of management, from the board of directors down to the shop floor supervisors. Each person, irrespective of his or her position, has a responsibility to try to meet the goal of the organisation. Buchanan and Huczynski (1985) consider that 'organisations are concerned with performance in pursuit of their goals. The performance of an organisation as a whole determines its survival'. Deming (1994) when discussing the role of managers related to the goal of the organisation offers the following: 'a good question for anybody in business to ask is *"what business are you in?"* '. He goes on to suggest that knowing the answer to this question is insufficient. It must be followed by a second question *'what product or service would help our customers more?'*. Deming's view of the role of management is clear, 'it is management's responsibility to look ahead, change the product, keep the plant in operation'. It seems such a pity that more managers do not subscribe to this opinion. Deming's view also reinforces the need to ensure that failing to meet the necessary conditions of a successful project does not violate the necessary conditions of a successful organisation. This is based however on the assumption that the strategies and, more importantly, the measures are properly aligned.

The role of measures in the current environment

In order to determine whether the goal is being achieved or not most owners will apply a set of measures. This might seem self-evident but in some companies that we have been working with, the initial impression is that the measures are not designed to determine progress towards the goal. Within most project-based industries these are usually net profit, return on investment and cash flow. At the end of each period the accounts are examined to determine the performance of the organisation and to see if improvement has taken place. This focus on the measures is a reflection of the goal, as expressed by Goldratt and Cox (1984), as being to make more money now, and in the future, through sales. If the goal is to make money then the measures must not only give indications as to how well the organisation is doing versus the goal, but also enable people to make decisions that are in line with the achievement of that goal.

However, the three measures of profit, return on investment and cash flow are usually referred to as global measures with little or no relevance to the supervisor, operator, team leader, project planner and first line project managers. There is a need for local measures to enable people to determine the impact of their decisions on the global performance of the company. The local measures proposed by Goldratt (1988) are throughput, defined as sales revenue minus the cost of raw material within a time period; investment, defined as the cash tied up in the business; and operating expense, the money needed to turn investment into throughput. Perhaps the best overview of the use of measures and comparing the Theory of Constraints (TOC)-based measures to the normal cost-based measures is provided by Noreen, Smith and Mackey

(1995) in their study of TOC and management accounting, and more recently by Corbett (1998) and Smith (2000).

Although measures figure highly in this book as a driver for behaviour, it is not the place to delve into the delights of accounting practice, whether it is cost accounting, management accounting or throughput accounting. There is however a need for a clear debate between a cost focus and a throughput focus in the area of decision making. I take the simple view that cost-based measures have lost their relevance in the environment of today. I consider that the TOC focus of throughput, investment and operating expense, and in that order for levels of importance, offers a far superior way to run a business with specific reference to the activity of making decisions. GAAP (generally accepted accounting practice) is still a requirement for the proper reporting of financial activity. The premise put forward by writers and researchers such as Smith and Corbett is simply that the role of GAAP is, and will remain, a legal requirement in the area of financial reporting. For the purpose of making decisions however, the far superior approach of throughput accounting makes the strongest case. Smith (2000) argues that

> The purpose of GAAP is to present a report of the past in a consistent and fair format. GAAP reporting records historical events and mathematically assigns a monetary value to the activity that has already taken place. The theory behind financial accounting is valid for the purpose of reporting past activities; however those actions necessary to maximise throughput and cash flow now and in the future are not the same as minimising local unit cost and maximising short-run reported net income.

The basis for this opinion lies in the way in which I feel a business should be viewed. We have, for many years, argued that the organisation should be seen as a chain of interdependent links. This requires that all measures, work practices and reporting procedures need to be aligned and also reinforce the behaviour the company is trying to seek from all of the workforce. This requirement is the area of greatest need for full and proper alignment between resources, functions, etc. throughout the whole revenue chain. It has important, and obvious implications for the financial measurement aspect.

Cost-based measures assume that each link is independent. If I have an activity where it is possible to gain substantial improvement in capacity then the decision centres on the additional capacity that is created at that resource area. This in turn might lead to a reduction in the cost of usage of that resource and thus be seen as a valid decision. Essentially what I have done is to add weight to a link. The assumption is that I have strengthened the whole as a result. Equally, if I am under pressure to reduce my cost base then once more I will seek reductions across every link. Each link will now be subjected to pressure to lose weight with often severe penalties in place for non-compliance.

But if the organisation is seen as a chain, and the notion of the weakest link is also recognised, then adding weight to already strong links does nothing for improving performance. Equally, cutting weight has little impact on the strong links and a devastating impact if the weakest link is cut. As those proposing cost-based measures lack the five focusing steps of TOC, usually described as a process of on-going improvement, they have no idea where to cut, therefore are almost certain to attack the wrong point. There are other implications and observations that could be made, suffice to add that a focus on throughput, linked to the notion of the weakest link, or constraint, offers a far better way forward, and a safer one than that of traditional cost-based analysis.

The weakest link determines the overall performance of the whole chain, it is the only place to focus improvement initiatives, it is where the decisions about which projects to undertake can be made, and it is the only place where real leverage takes place.

Smith (2000) states that

Agreement on the need for a measurement system that encourages local actions in line with a company's bottom line results is common but it continues to be elusive. Traditional cost accounting product cost allocation, using direct labor to assign overhead to the products with a focus on local labor and resource efficiencies is at best irrelevant and at worst dysfunctional. . . . Concepts such as activity-based costing, then activity-based management have evolved to answer the deficiencies of absorption unit costing for decision making.

She concludes that

As commonly implemented they all create local optimization that puts cost centers in competition for resources and results in conflicting actions between departments. They all ignore the impact of the scarce resource, the constraint.

The role of change

If addressing the issues of financial performance and the other business issues described earlier were all that concerned managers, although appearing difficult, it would not be impossible. However there is also the fact that radical change is taking place across the whole spectrum of business life. Change in terms of technology, the market expectations, the levels of quality and so on are a constant theme in most sectors of an economy. Likert (1967) argued

> *Every organisation is in a continuous state of change. Sometimes the changes are great, sometimes small, but change is always taking place. The conditions requiring these changes arise from both within and without. As a consequence there is a never-ending need for decisions, which guide adjustments to change. The adequacy of these decisions for meeting an organisation's current and developing internal and external situations determines the well-being, power and future of that organisation.*

Skinner (1974) in describing the focused factory suggests that

> *Focused manufacturing must be derived from an explicitly defined corporate strategy which has its roots in a corporate marketing plan. Therefore the choice of focus cannot be made independently by production people. Instead it has to be a result of a comprehensive analysis of the company's resources, strengths and weaknesses, position in the industry, assessment of competitors' moves, and forecast of future customer motives and behaviour.*

If the statement expressed by Skinner is accepted, then many of the rules and measures, etc. used in most organisations are erroneous, which I define as being those leading to unwanted outcomes in terms of either financial performance or individual or group performance; then when linked to the notion that measurements should motivate the people within the various parts of the organisation to do what is best for the organisation as a whole, some measurements do not achieve that purpose at all. They do not induce people to do what is good for the organisation. At the same time, if training conditions people, and there is a lot of training taking place, and erroneous factors already described exist, then it is very likely that many people are conditioned to follow rules, even erroneous ones. Of course, the intuition most people have is very strong so there is the real possibility that many people are, at least subconsciously, disagreeing with their own actions. If people are rewarded according to their performance as measured, and these same measurements are also erroneous, then people are forced to behave in line with some erroneous measurements and often find themselves valued somewhat arbitrarily.

This leads to two conclusions. The first is that if people find themselves at odds with their intuition, and yet are forced to behave that way, then they find that they are in conflict with common sense. Also if they feel

9

valued in that way then some will feel that they are undervalued while others are overvalued. Given that satisfaction is important to people then frustration is a normal outcome of a discrepancy in evaluation. As this frustration then requires an outlet, people start to fingerpoint and the blame culture becomes apparent, or some people just give up and become apathetic. If there is a real sense of injustice then the likelihood is that there will be considerable activity behind the scenes and political manoeuvring. This in turn leads to a sense of protecting one's back and the whole scenario leads to walls of distrust between levels and functions.

The outcome of this rigid application of erroneous measures and rules, coupled with training and education based on erroneous assumptions, is a company that is facing real difficulties and certainly has no real chance of improving performance in the long term. Tichy (1983) has proposed one response to this problem. He considers that

> *The response to managing in turbulent times requires organisations to return to basic questions about their nature and purposes. The fundamental character of their technical system will need re-examination resulting in new missions and strategies, major re-structuring and revamping of the financial, marketing, production, and human resource systems. Organisations' political systems as reflected in who gets ahead, how they get rewarded, and who has power to make decisions would also need overhaul. Organisations' cultures are perhaps the most complex and subtle yet most pervasive on their effectiveness. Thus major change will require addressing issues of values and beliefs of organisation members.*

Tichy is referring to the need in times of economic and political turbulence for the ability to continually examine the direction and nature of the organisation. This must include a review of the basic assumptions that led to the formation of the organisation in the first place. In turn, this should cover the technical, socio–technical and social systems of the company and include all the people within it. The review ought to go right to the heart of the organisation and challenge these basic assumptions. It also requires the examination of the many causalities that exist, teasing out the real and the erroneous, and starting the organisation once more on a process of on-going improvement.

This final statement, concerning on-going improvement, is found throughout the writings of people such as Goldratt (1990), Deming (1986), Imai (1986) and Feigenbaum (1991). They each recognised the importance of ensuring that performance is continually reviewed and upgraded. For them any organisation that does not improve is already slipping behind the competition, hence the continuing need to analyse performance, seek out the areas of non-performance and deal with them properly. This is the starting point for such approaches as the 14 points of Deming, the statistical approaches of Feigenbaum, the Kaizen described by Imai and the TOC of Goldratt.

One final thought in this debate is summed up by Deming (1994). He states clearly that 'apparent performance is actually attributable to the system that the individual works in, not to the individual himself'. Which is why Deming, I and many others have strongly argued that the current practice of ranking performance of individuals and even depart-ments is both a farce and highly divise. Of course, this in itself could be the subject of a whole book, suffice to say here that the application of measures should be to determine whether actions being taken are leading to enhanced performance or not. If they are being seen as divisive then the expected benefits are not being achieved.

Introducing the TOC dimension

The Theory of Constraints (TOC) is a process of analysis and problem solving with a history primarily in manufacturing. More recently this has expanded, as a result of the development of the TOC solution in project management, to a far wider range of companies in areas beyond manu-facturing. At the heart of any TOC analysis is the realisation that without an holistic approach the results are at best local, and often in conflict with the overall strategy of the organisation.

The debate so far has shown that there are two key elements in determining the operational success of an organisation. Assuming that the goal has been defined, and this is not always the case, and the measures appropriate to the goal have also been determined, then progress towards the goal can be measured. Fine so far, however, from the debate two clear, distinct and potentially conflicting choices appear. The first is what might be described as the *corporate focus* where cost and the issues of cost allocation and profit allocation are para-mount. The second side is very much on satisfying the needs of the market, the *market focus*. In the current environment these two aspects are seen as choices. Examples abound where managers' bounce between the two choices depending on the pressures being brought to bear on them. At any one moment in time one will dominate to the almost exclu-sion of the other, and then switch back again. Within TOC this type of conflict is defined as a *cloud*. Throughout this book clouds will be a common feature as they represent one of the most effective ways of describing conflicts and, having described such a conflict, in determining how to remove it once and for all. This whole process is known as *evaporating clouds*.

Clouds and the creation of understanding

The term *evaporating clouds* is drawn from the book by Richard Bach entitled *Illusions*. The phrase used within the book is actually *vaporising clouds* which to my mind is more positive than simply evaporating. Vaporising clouds seems a more proactive activity, which demands positive action. The structure of the basic cloud is shown in Fig. 1.2.

The main boxes, entitled **A**, **B**, **C**, **D** and **D′**, make up the basic structure of the cloud. The logic of the cloud is that of necessity. Thus **A** is the objective of the cloud, what those involved are trying to achieve as an overall objective, and both **B** and **C** are necessary conditions for the achievement of the objective as written within the **A** box. Although defined as necessary conditions, they are not always sufficient for the objective to be achieved. In the same manner **D** is the necessary condition for **B** and likewise **D′** for **C**. The conflict exists between the **D** and **D′** boxes. Once the cloud has been verbalised in this way the next stage is to ensure

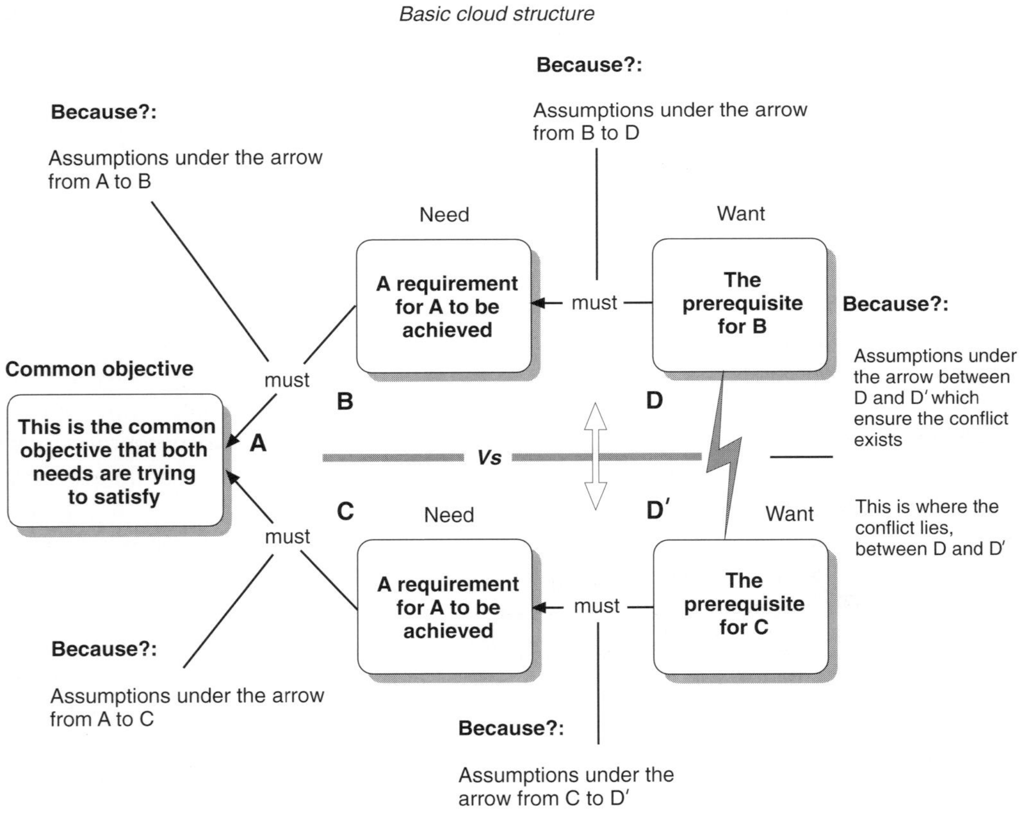

Basic cloud structure

Because?:

Assumptions under the arrow
from B to D

Because?:

Assumptions under the arrow
from A to B

Need   Want

**A requirement
for A to be
achieved**   ← must —   **The
prerequisite
for B**   **Because?:**

Assumptions under
the arrow between
D and D′ which
ensure the conflict
exists

Common objective   must

B   **D**

**This is the common
objective that both
needs are trying
to satisfy**   ► **A**   ═══════ **Vs** ═══════

C   Need   **D′**   Want   This is where the
conflict lies,
between D and D′

must

**A requirement
for A to be
achieved**   ← must —   **The
prerequisite
for C**

Because?:

Assumptions under the arrow
from A to C

Because?:

Assumptions under the
arrow from C to D′

Fig. 1.2. The cloud diagram

that the logic holds true by reading the cloud and amending the verbalisation where necessary. The cloud is read from the tip of the arrow to the tail. For example, 'in order to have (A) I must have (B); in order to have (B) I must have (D)'. The rest of the cloud is read in the same manner. Once the verbalisation of the cloud is clear, the next step is to surface the assumptions that lie behind every arrow. This is done by adding the word 'because' after the phrases above. For example, 'In order to have (C) I must have (D') because....' The assumptions that are surfaced are then placed in the appropriate box and then scrutinised to see if they are true, valid or erroneous. If an assumption is found to be erroneous then it may be possible to break the seemingly unbreakable conflict that exists. In this way the objective can be reached. For a full description of the construction of clouds see Goldratt (1990) and Scheinkopf (1999).

Conflicts can take many forms and therefore clouds can represent a number of different types of conflict. So far the discussion has represented a choice between two competing approaches to financial management, the use of cost as the focus versus the use of throughput. This results in a choice cloud (Fig. 1.3).

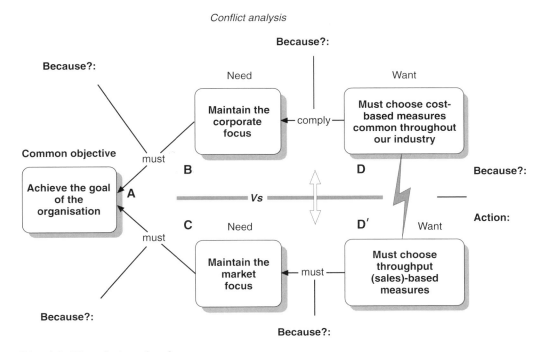

Conflict analysis

Fig. 1.3. The choice cloud

If we read the cloud properly then this is what it says.

> *In order to achieve the goal of the organisation we must maintain our corporate focus. Also in order to achieve the goal of the organisation we must maintain our market focus. In order to maintain our corporate focus we must use the cost-based measures that are common throughout our industry. However in order to maintain our market focus we must use throughput (sales)-based measures.*

Now constructing the cloud and subjecting it to the initial scrutiny is not enough, we must also examine it to see how powerful the conflict is between what is written in the **D** box and what is written in the **D'** box. This can be done by checking the impact of **D** on what is written in **C** and the impact of **D'** on the **B** box. If we start with the **D** on **C** connection, what is clear is that many companies, in using cost-based measures, actually destroy their ability to grow sales. Decisions about allocations lead to some products being dropped and some projects not being delivered in line with market expectations. The resulting impact is on the market's view of the company. Equally, if we focus on the throughput-based measures, those doing so quickly fall foul of the corporate measurement system and its insistence on corporate measures. Efficiencies will not be as expected, resources will not meet the targets set for them, and so on.

This being a choice cloud it is common practice to see people choose which side they are going to operate on. Sometimes they find themselves bouncing between the two. The cloud encapsulates the chronic problems facing companies today with the ever-increasing pressure to focus on short-term solutions. Deming (1994) offers this view of the conflict when he writes

> *No number of successes in short-term problems will ensure long-term success. Short-term solutions have long-term effects. Of course management must work on short-term problems as they come up. But it is fatal to work exclusively on short-term problems, only stamping out fires.*

He might usefully have added the activity of lighting a few fires for tomorrow just in case as another example of problematic behaviour.

Part of my research identified what I call a hierarchy of clouds, a set of four clouds, which connect in a specific manner. At the top of this hierarchy is the choice cloud, the next level is the decision cloud, below that the conflict of subordination cloud and, finally, the paradigm lock cloud (Fig. 1.4).

Once a choice has been adopted, the decisions that are taken are reflected in that choice. Therefore once a choice has been made within the context of the choice cloud, all subsequent decisions are driven by

Fig. 1.4. The hierarchy of clouds

that choice. However the decisions themselves may also contain inherent conflicts – hence decision clouds. Thus, in the taking of decisions conflicts once more appear. Driven by the choice made earlier these clouds operate at a lower level but have just the same level of impact. Figure 1.5 is an example of just such a decision cloud. The measures in Fig. 1.5 are related

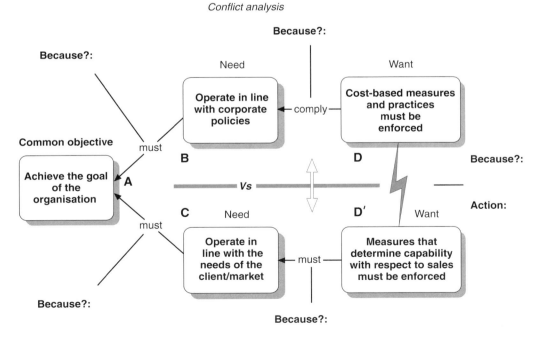

Fig. 1.5. The decision cloud

to the choice made earlier. Within a cost-based measurement system there are still delivery measures, quality measures and time measures. The decisions however are conditioned by the need to focus primarily on cost. All these other measurements must fit into such an overall format.

Having taken the decisions there is the third in this hierarchy of clouds – the conflict of subordination cloud. Decisions are suppose to be followed, but what happens when following the decisions makes no sense in the mind of the person responsible for implementing the decision? Here is a real conflict once more – that of subordination. Often people in the various programmes we have run throughout Europe observe that they have to implement cost-based, efficiency measures. However once they have worked through an analysis of their company, and the way in which it generates money, they switch to the other side of the original choice cloud and feel the strong need to work according to sales/throughput measures. At this point the pressure to conform to the required financial measurement system becomes fierce and they find themselves between the proverbial rock and a hard place. To the people trying to make the system work this only confirms the inadequacy of senior management to really focus, and set appropriate priorities on what is important, the generation of money and not the saving of it.

The questions that need to be answered

Given all the available approaches to improvement in popular usage one would expect to find many organisations experiencing rapid and sustained growth. The reality is that many organisations find that the expected improvements do not appear. Although most companies have a plethora of improvement projects underway at any one time; although there is substantial investment in new tools and technologies for project management; and although most companies fervently seek newer and better ways of making money, few seem to be successful. This applies even when the proposed solution is fully developed in line with the analysis of the current situation and is developed with the full support of the interested parties.

This failure suggests that there is a further obstacle to improvement not defined in the usual adopted methods. If it is accepted that most, if not all, improvements involve change, often significant change, then the combination of a need to improve and the requirement to change suggests that the obstacle to change lies with the people involved, irrespective of their level of involvement. If this is the case then it can be argued that, given the need for change, the actual behaviour appears dysfunctional. The observable effect, which prompts this problem, is that of a dysfunctional behavioural constraint preventing an individual from implementing a solution. The question therefore centres on the nature of this constraint. Can it be determined, described, and if so can it be done in such a way as to both illuminate and suggest a way to alleviate the impact of this constraint? It was in trying to answer this question that both this book and the research that lies behind it was first undertaken.

I believe that there are two distinct dimensions to the answer to this set of questions. The first is organisational and the second personal. I argue that the primary reason why organisations find it so difficult to sustain improvements over time is that they overlook the impact of the constraint. In trying to maximise performance in each of the key performance indicators outlined earlier, they manage the business without any reference to what I call the *laws of constraint management*. The personal dimension is the paradigm from which the individual works. In order to achieve the kind of success sought by many people both of these dimensions have to be addressed. The laws of constraint management demand an enterprise perspective for all within the organisation. This requires, in many cases, a change in the behaviour of the people within the company. If this change challenges a paradigm then the importance of change management becomes key.

Therefore for a project-based organisation to develop an enterprise-focused approach to managing projects the importance of the constraint has to be addressed, and the importance of taking the people with you has also to be addressed. It is these two issues that the rest of the book addresses.

One final thought for this chapter. To my mind there is only one key constraint in any organisation. If the intention is to substantially increase throughput I have discovered only one genuine constraint. It is not the market, it is not capacity, it is not the size of the warehouse, it is not a lack of people, and it is not a lack of money. In fact it is not the lack of anything except one dimension that only human beings can bring. Sure the rest all seem to be potential constraints. To my mind they are only obstacles on the path to the goal. They can all be overcome if, and it is a big if, we unleash the intuition of our people. The only constraint operating within all of the companies I have worked in over the years has been, to my mind, a serious lack of imagination being allowed to develop. This book, the research that lies behind it, the TOC/TP tools are all about allowing people, from boardroom to shopfloor to think, to release the energy of the mind.

CHAPTER TWO

The role of the Theory of Constraints within the enterprise business development of project-based organisations

Introduction

This chapter focuses on the role that the Theory of Constraints (TOC) has within enterprise-focused, project-based organisations. The starting point for the chapter is the importance of setting the project environment into the context of making money now and in the future – the goal of the organisation. This level of clarity allows for the ability to determine progress towards the goal. This assumes that the goal has been properly communicated throughout the whole of the organisation, any necessary conditions also communicated and measurements in place to determine progress.

If in our case the goal is to make money, and the process by which the goal will be achieved is through the completion of projects, then how are we doing? Given that there is frequently a gap between expectation and reality this begs the second question, why is it so difficult to make money from projects? This leads us to the discussion about the application of the theory of constraints, the nature of constraints and how they link back into the subject matter of projects. The management methodology of the TOC is rooted in the analytical ability of the thinking processes (TP) to determine core problems, core drivers of the issues that are taking up so much management time and energy. The TP tool of most use here is that of the *evaporating clouds*, usually abbreviated to simply *clouds* and introduced in the previous chapter. This then leads us to a discussion about the central laws of constraints that affect all organisations, whether they recognise that fact or not.

An overview of the TOC process and organisational analysis

The application of the TOC process to the analysis of organisations grew out of a sense of frustration that piecemeal analysis led to no real improvement on the bottom line. This is based on a number of key assumptions. The first is that key business indicators such as shareholder value and market share are all giving clear signals as to how the organisation is

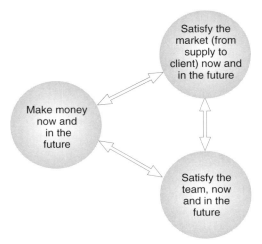

Fig. 2.1. Linking the goal and the necessary conditions

delivering compared to the goal set for it, the goal being defined as the ability of the system to make money now and in the future. Achieving the goal is dependent on at least two necessary conditions. The first is in satisfying the market, from supply base right through to the client. The second is in satisfying the team, the team being defined as the group of people who operate together in order that the system works effectively. This is often represented as in Fig. 2.1.

I have made reference to this as a system. I like to think of it as a revenue system, but what is more important is the way the system operates. There are two conflicting views of how the system operates. The first sees each element within the system as independent and the second as interdependent. The TOC perspective, the choice made by those who implement and manage TOC-based solutions is that of the interdependence of each link in the revenue chain. The revenue chain is the model used within the TOC environment, stretching from the supply base right through to the client to determine the flow of product or service and money. Thus if the analogy of the chain as a revenue system is accepted, and if the strength of that chain is determined by the weakest link (the constraint) is also accepted, then the most effective way to manage such a system is to focus on that constraint. The focus of the management must be to analyse, manage and improve the overall performance of the chain through the ability to manage the constraint.

The chain consists of the various functions/resources within the organisation. Although shown here in a linear fashion (Fig. 2.2) it will rarely be this simple. The important aspect to remember is that all the functions within the organisation need to be included, and that means the support

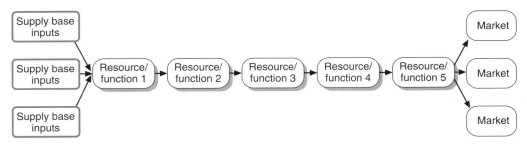

Fig. 2.2. The revenue chain

functions of finance, sales, marketing, administration and so on, as well as the more obvious line function people such as production, or the project teams.

Before we consider how to analyse the performance of the chain let us consider how to measure performance in the first place. In chapter 1 I drew attention to the work being carried out by Smith, Corbett and others into the wider acceptance of throughput accounting (TA) for decision making. Here I want to place throughput, and the other key measures of TA into context. Profit, return on investment (ROI) and cash flow are rightly recognised as the key measures in any for-profit organisation. However it is usually very difficult to examine the impact a decision might have on these measures except at the very highest level. Other measures are needed in order to give both focus and confidence that the decisions being taken will move the organisation closer to the chosen goal. TOC proposes that three measures should be used within our revenue system: throughput (T), the rate at which the system generates money through sales (T = sales revenue less true variable cost); investment (I), the level of money tied up within the organisation; and operating expense (OE), the amount of money required to turn investment into throughput. These three measures: T, I and OE form the core of the TA system of financial measurements. For more on this aspect see Goldratt (1986), Srikanth and Robertson (1995), Corbett (1999) and Swain and Bell (1999).

Within TOC the link to the three primary measures is made by the following equations: profit is T − OE; ROI is (T − OE)/I; and productivity is T/OE. As these measures can be checked weekly this gives an excellent way of measuring the performance of the system. Now, when making a decision and taking note of the constraint, questions can be asked of any decision. Does it increase T? Does it reduce I? Does it reduce OE? Reducing OE is valid if the activity does not impact the ability to generate T. Equally, seeking more T only makes sense when the costs do not outweigh the increase in T that is expected. It is this approach to measurement that will feature heavily in both the cases studies and in the proposals contained within this book.

Setting project management into the context of achieving the goal

Those companies that expect to generate revenue through projects do not do so out of some desire to act more as charities. They do so with the specific aim of producing what the client wants more quickly, more effectively and at a lower cost than the competition. This then governs the choices they make about their structures, and the methodologies they adopt in order to achieve the stated goal. It will determine the tools they implement to deliver their projects, and the methods by which they address their chosen market or markets. If the goal is related to the accrual of money over time then the ability to deliver projects on time is a key issue. There might even be additional advantage, and value in delivering the project ahead of time. Note that there always seems to be penalties for late delivery. Therefore managing projects with respect to the due date is fundamental to a successful operation. The focus throughout the project should therefore also be progress towards the due date of the client. All of the decisions now taken are subject to that choice – meet the due date of the client. All of the planning decisions should support that decision. Indeed the work breakdown structure and the subsequent scheduling of the project should all be done in line with the achievement of the due date.

But time is not the only indicator of success in aligning the goal and the project. Clients expect to see projects completed according to their specifications. They do not expect to have bits missing. If the specification asks for four lanes of highway there is an expectation that that is what will be built. However if some of the actual experiences of project-based organisations in the high-tech environment were to be transferred to civil engineering we would find new roads being built with two lanes on one side and only one on the other. The whole specification is part of the deal. The client does not expect to see corners being cut to satisfy some other requirement. Thus in order to meet the goal, the specification, the whole specification, and nothing but the whole specification of the project must be delivered.

There is one more necessary condition in the linkage between the goal and the successful delivery of projects and that is budget. It is a fair assumption that there is no bottomless pit of money for the project team to dip into. Therefore the budget must be complied with which in turn places an important responsibility on those making decisions throughout the project to make them in line with both the deliverables and the budget. There will always be times when decisions related to budget have to be made. The question is how good are the information systems to give the proper data on which to make sound decisions?

The assumption within this book is that only when projects are delivered on or before time, and when the client specification is met in full, and when the budget is complied with, can the project be termed successful and progress toward the goal of the organisation enhanced. Therefore all the aspects of running successful projects must meet these criteria. They are the strategic choices made at the outset. All decisions are now dependent on those choices having been made and agreed upon by all concerned. The revenue chain is dependent on the proper alignment of all three criteria. These are now the necessary conditions for the maximisation of the revenue chain, from supply to client. What remains at this point is the recognition that the organisation itself must be aligned with the criteria for a successful project.

Some key problem areas in generating revenue from projects

Having determined what constitutes a successful project – meeting the three criteria of time, specification and budget – the next question is 'do all companies meet those targets every time?'. Of course it is clear that there are many companies who do succeed in making money from projects, but the question remains, is it enough? Do they make the most out of the projects that are available to them? Could they, with the same level of resources, deliver more projects within the same timescale? What are the problems that already prevent them from achieving the levels of success they currently seek, never mind what might be set as targets for tomorrow? How many times are project meetings given over to a discussion about how to apportion blame, and in some cases try to make the client pay for the delays, which have resulted? How many times do the discussions revolve around methods to try to reduce cost, often without reference to the damage that that course of action might have on both the client and future throughput? It is often easy to blame the client who has been slow to provide clear specifications, the weather, the inherent risks within the technology, the unreliable nature of the subcontractors, the unforeseen emergencies that happen every time! All in all it is the high level of uncertainty that every project has which leads inevitably to the problems, and if you listen to some project managers, this level of uncertainty is simply an act of God for which nothing can be done.

A simple case study drawn from many possible sources but containing the core elements of what we are discussing might help. It is, of course, a work of complete fiction! Tom is a project manager. He has about 20 people in his department and he has three main projects underway at any one time. People are spread across all projects and the recent history suggests that he is in difficulty. He has not hit one single project due date for nine months. Senior management have noticed and Tom's boss, Cathy, has told him to get his act together and start delivering. He has asked for more staff only to be told there is no budget. He has to deliver with what he has. At the same time the corporate people have been looking at the way he is using his people and have found that often they are not working on projects when they should be. He has been told in no uncertain terms that he has to increase the productivity of his department or someone else will. Yesterday Cathy called him into her office to announce that she was going to give him a new major project. This project would give him the chance to demonstrate his capability, and would ensure that the division remained a core part of the organisation. The project was high profile, about 18 months of work, and was in the latest technology. She wanted to see the project plan on her desk within three weeks. To create this plan, with over 200 engineers involved, and with over three sites spread across Europe, was going to be no easy task.

Tom sets off to start planning the project. He knows from bitter experience the impact that uncertainty can have on projects. Nearly all of his past and present problems have stemmed from the levels of uncertainty that have occurred during the lifetime of his projects. So he starts with a simple aim, he wants to manage the levels of uncertainty better than ever before. At the same time he also knows that the overall lead-time of the project is crucial to the success of the project. Therefore he must deliver the project on time while at the same time dealing with the uncertainty that is certain to happen.

Now the overall lead-time of the project is determined by the sum of the task times for each activity on the critical path, and that lead-time has to match with the due date already announced. Already there is pressure to keep these critical task times to a minimum and he has announced this to the project team members already, with little enthusiasm coming back from them. At the same time Tom has learned that it is only during the execution phase of the project that the real level, and location of the uncertainty, is known. Uncertainty is usually translated into accounting for this at the estimation phase with each activity having an element of protection added in, the float.

There is little point in having task times if they do not reflect the reality of the situation and that means having task times, which can be given some form of guarantee. But here comes the rub, how is it possible to have a set of task times that reflect the urgency with which the project must be finished and at the same time have enough protection built in

23

to protect the project itself. Tom remembers the many meetings to discuss this very point, and how ugly they became.

Tom is only too aware of the many times that in order to meet the due date of the projects, he himself has cut the times given to him by the planners, and how he has then found that the planners, and the engineers, have added time to their estimates to counter his cutting the times. If that were all he had to contend with it would be bad, but then once the turmoil and anguish of the estimation process has ended, and the project has moved to execution, he has had to contend with people working on other projects at the same time, thus delaying starting on his project, or leaving his project due to some crisis elsewhere and then taking ages to come back to his work, and in some cases people just not starting when they said they would citing that there was plenty of time anyway! All he ever sees is that the protection time, the float, is being used up, and in some cases has been completely used up, long before even the halfway point of projects.

Of course everyone can point to the disasters that have occurred. They all have chapter and verse as to why the problems have occurred and why little can be done about them. That is why it is called uncertainty. Which is why whenever Tom and the resource managers are meeting to discuss projects they are convinced that the level of uncertainty is increasing each and every time. The number of tasks that are late just seems to increase and every time the project seems to require rescheduling, with the critical path often changing two or three times a month, or even a week as in one or two recent cases. This constant updating of the plans seems to take forever. The new schedules are often invalid by the time they are released, and in some cases there is no time left to recover and therefore the whole project runs late. Once more Tom is being hauled over the coals.

When he has tried to expedite by using outside contractors, the additional costs have come out of his budget, and once more the wrath of the finance officer has been brought down on his head. Equally, when he has suggested that the specification be cut in order to meet the time schedule of the client, marketing has sent around some people to remind him of what he has promised to deliver. All in all, a very difficult situation to manage, and one in which Tom is becoming more and more convinced that there has to be a better way.

Some basic dimensions to projects and project management

Let us look more closely at the predicament in which Tom finds himself. Our research has shown that this is not a unique situation, that the story painted here is repeated time and time again in many companies around the world. The key issues seem to centre around the question of uncertainty. There are some basic dimensions to any project. One key issue relates to the whole question of variation. Tasks have variation built into them. The technical content can prove to be difficult thus causing delay in completion. The people often find that the time they gave as an estimate is insufficient to meet the actual technical content of the activity. There are many reasons why the actual time to complete differs from that estimated, and is often longer. Therefore, one of the key responsibilities of project managers is to reduce the level of uncertainty wherever they can, and at all times manage the level of uncertainty encountered to minimise violation of the due date. In managing uncertainty more effectively, project managers feel more at ease.

The problem lies in the fact that it is only when the project moves to execution that the actual delay becomes apparent. The degree of uncertainty, and the effect it is likely to have on the overall project, only becomes clear once the reporting system flags it up. Couple this with the fact that the time to complete the project determines the revenue generation of the organisation and we find that a set of severe problems are now being encountered. The first comes in the form of a response to the possibility of meeting such variation, and the tendency is to include the possibility in the estimation. It is often the case that the actual time estimated is not the time to complete the task, but the time to complete the task with a factor for variation built in. Remember giving an estimate is giving a commitment in many companies, and people being asked for estimations are careful not to commit too much too early. Hence the estimated time is not the actual time to complete but a time that takes note of the degree of uncertainty. The process of setting task times in this fashion is based around the assumption that the best place to deal with the variation is at the point of each task. Where better to deal with the uncertainty of a task but at the task itself. This leads to task times which are longer than might otherwise be the case, but have the advantage that they are each capable of dealing with the level of expected uncertainty.

But let us not lose sight of the fact that project completion is the point at which revenue is generated. If the task times are longer just to ensure confidence about dealing with possible variation, then the overall project time is also longer. This means that the due date is further out in time than might otherwise be the case. This might have important commercial implications. The need to meet the critical time factors of the project is also uppermost in the minds of the project managers. Now they are faced with a dilemma. On the one hand they are under pressure to safeguard against variation that might exist within each activity of the project, and at the same time they are under pressure to meet the needs of the

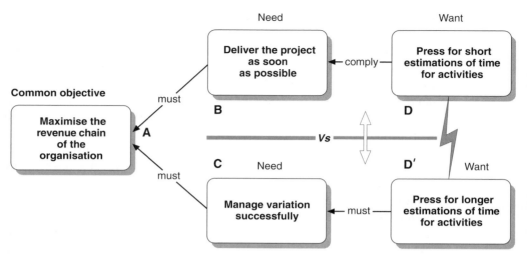

Fig. 2.3. The conflict of estimations

customer and deliver the project as soon as possible. This is a classic conflict cloud as shown in Fig. 2.3.

This cloud leads to many fights in the planning meetings, in the senior management meetings, and in meetings with the client. There is usually no dispute between the parties about the objective, maximising revenue, or about the need to both manage variation successfully and deliver the project as soon as possible, and certainly by the agreed due date with the client. The issues are about how to achieve both of these seemingly incompatible conditions. At these meetings conflicting agreements are often reached in order to pacify one side or the other. The bottom line remains the same; the project is almost certainly under severe pressure from the outset.

For example, if the overall project plan gives a due date which is after the expected delivery to the client, senior management will often arbitrarily cut the times. They do this not because they are trying to impose their will on the project managers and engineers who created the project plan, but because they recognise the need to maintain a focus on the market and the customer. When this is coupled with the demand to keep the task times high, the inevitable outcome is that the engineers also factor in time which they expect to be cut, but which after the cut has taken place still has enough safety time to cope with the potential level of variation. This game in the setting of time estimates can often go on for quite some while. Now there is in addition, the real problem of trust between the two parties.

Estimation is one thing; execution is quite different. Once the project moves to execution the people doing the tasks know about the additional time elements which have been factored in for variation. They know that there is time to do the activity and they can be lured into delay. They might not start an activity for a variety of reasons. They might have other project tasks to complete first, on both this project and, more likely, other projects. They also have to contend with the fact that the current project might also be already suffering from delays itself.

Finally, there is a phenomenon known as *student syndrome* which can be summed up by saying that we can put off today what can easily be done tomorrow. We put the task off for a variety of reasons, perhaps because we do not really know how to do it, because it appears to be straightforward, that it is a task we have done often before without problem and so on. As all this is happening, and no work is being done on the task, time is slipping away. All the time for variation protection is used up, all of the time for possible problems has gone, and the result is that the remaining time is often only just enough to complete the task assuming no problems whatsoever, and no other tasks to be completed within the same time period. Of course it can be counted on that as soon as a task becomes critical something will happen to make it even more critical. Time is once more lost, except that by now there is no protection time left to protect, the project is eating into the bone, the fat having been consumed already.

Post hoc rationalisation now takes place, the engineers and the resource managers are now convinced that time should never have been cut by senior management. Project managers are now convinced that time has been lost due to the inability of the resource managers to focus properly and there are often fights and accusations flying around the project team meetings. This leads us to the same conclusion as Tom, there has to be a better way of managing projects. The TOC approach offers a way forward. Analyse the system of the revenue chain and try to determine what is preventing better performance from taking place. In other words find the constraint.

The TOC approach to dealing with constraints

The TOC approach to dealing with constraints centres around the five steps of focusing. These steps, first described by Goldratt (1986), give a clear process, which can, if used properly, take the company forward. The five steps are as follows

1. identify the constraint
2. exploit the constraint
3. subordinate to the constraint
4. elevate the constraint
5. if the constraint has been broken in the preceding steps, prevent inertia and go back to step 1.

These steps need further explanation if their real power is to be recognised. Identifying the constraint is the starting point, although it might be argued that if there is a search for a constraint then the trigger for that has to be some form of non-performance. From what we have discussed already it is clear that many project-based organisations are not achieving the levels of financial performance they should, and in some cases losing heavily in markets they should be dominating. Equally the constraint can operate at more than one level. For the organisation as a whole there is the likelihood of a constraint operating, and also at the project level a different constraint applies. Just as every organisation will have a constraint within the overall revenue chain, each project or set of projects will have a constraint operating within that local revenue chain. In some cases they might even be the same constraint. But are all constraints the same?

A typology of constraints

The research that has taken place over the last few years into the application of the TOC has revealed a typology of constraints.

Physical constraints are thought to be the most common. In reality they are the least common in the true sense of the term. Typical examples from a manufacturing environment could be a machine tool, physical space, number of workstations, etc. It might include the number of people, the number of hours worked, the availability of material. Within project environments it might be machinery, the number of test stations, the number of gangs employed and so on. However this last group could also be the result of a different form of constraint – a policy constraint.

Policy constraints are much more common than usually accepted. If we look carefully at the last three examples of the physical constraints, number of people, number of test stations, availability of machinery, each of these constraints which appear to be physical could just as easily be the result of policies within the organisation. In each case the goal of the policy is not to achieve the goal of the organisation in terms of making money, rather to meet the goal by not spending any money. The number of people being too few can be the result of the policy of not recruiting for a specific period of time, the number of hours might be a policy of not allowing overtime, not allowing a second shift and so on and the availability of material might be the result of only going to the cheapest supplier who regularly fails to meet the wanted delivery deadline of demand. In my experience, policy constraints are driven more by the rules and measurements that exist than any other contributing factor. In the many organisations I have been involved with over the past ten years I would argue that the split between physical and policy is about 20% physical and the remaining 80% being policy driven.

Policy constraints are often the cause of much of the misalignment between the apparently conflicting necessary conditions of projects and those of the overall organisation.

Paradigm constraints exist in all organisations with all people. They are a function of each individual and the way in which they view the world around them. More will be said about these constraints later in the book. Suffice it to add at this point that for each implementation described within the case studies, the paradigm constraint was perceived to be the most effective in delaying the benefits, and in some cases led to the project not achieving the proper targets at all. Paradigm constraints are, at the same time, the most difficult to deal with.

So, returning to the five steps of focusing, the next step is that of exploiting the constraint. Exploiting the constraint refers to the need to ensure that the most is being produced from the constraint area before any additional investment is made. It is often the practice that once the constraint is identified additional capacity is acquired straight away. The TOC approach argues that this is often counterproductive and that the steps of exploit and subordination should be done first. Exploit is about making what you have work better, more effectively than before. Is the resource really being used all the available time? Are the machines really working in alignment with the objectives of the project? Is inspection carried out before the constraint operation thus preventing faulty material going through the constraint? Is preventative maintenance being carried out as a scheduled activity thus ensuring substantially reduced loss of time due to breakdown or loss of specification on the machine?

Subordination is a tough call. In ten years of experience of implementing TOC solutions I find that this is the step most people find the hardest to accomplish. Much more will be said about this aspect later. Put simply, subordination means that everyone, every machine, every operation, in all functions, both line and support, focuses on the performance of the constraint area and keeps that focus irrespective of any other force. The key to subordination is the importance of ensuring that the rules and procedures for the new methods are clear and unequivocal and that they are monitored from the highest level. Also that the reasons for subordination are communicated clearly and effectively, with all participants able to make their reservations known if the outcome is to be at all successful.

Elevate is simply to add capacity to the constraint. This can have the effect that the constraint moves, what was the constraint is no longer the constraint and therefore a new constraint appears. Therefore the fifth step is return to step one. This is all about inertia. Preventing inertia is about having once overcome a constraint, do not be lured into the trap of continuing to elevate an already elevated constraint. The logic of the process determines that there will be a new constraint and we have to go round the loop once more, thus ensuring that we are always seeking improvement in performance.

The links to project-based organisations and to projects themselves

Applying the five steps of focusing within the project environment, the first two steps, identify and exploit, are clearly involved. There are physical and policy constraints in all project-based organisations and in all projects. Before examining the project-based constraints let us consider a typology of projects. There is clearly more than one type of project. Tom is working in what is known within TOC as a multi-project environment. This is where key people are spread across more than one project within any period of time. It can also be where key equipment is also spread across more than one project in the same timeframe. Multi-project environments offer the most confused and complex areas in which to work. Confused because the rules of working often appear to be conflicting dependent on who is late, which customer just called, which manager is behind on his local performance measurements. Complex because the tools used to describe such environments fail to really capture the true picture and also fail to give clarity for making decisions. This is why the usual practice is to multi-task between project activities. The next type of project is multiple single project. This is where the organisation has a number of projects on-going at any one time but the resources are not scheduled to do more than one project activity in any one timeframe. They can still be working across projects, it is the timeframe that is key here and there is little need to multi-task. People are working on individual projects independently of all the other projects that might be going on around them. The final set is that of single project environments. This is where there is only the one project, and people and the organisation have to finish that project before the next starts.

For the case studies contained within this book the primary examples are multi and multiple single project environments. The next step is to identify the constraint. To accomplish this task, given the fuzzy nature of the problem area, it is necessary to use the analytical tools of the TOC/TP. This is to ensure that we have identified the true core problem(s) and gain real consensus on the problems of projects and project management.

The analytical process of the TP

This section of the book describes the process used to analyse in more detail the problems, which affect both projects and the organisations associated with them.

The starting point of any TOC analysis is the set of problems being experienced by the people, the managers, the engineers and so on, who are involved in the area of focus. The problems are often stated in fuzzy, almost bland language, and the first step is to gain clarity of each problem statement. For example, one often used statement is 'communication is poor'. But what does this really mean? Is it financially poor? Is it an example of destitute communication? What is meant by communication? Is it verbal, written or in some other medium? The statement fails one of the reservations used within the TOC logical thinking processes known as *clarity*. Clarity is not really a logical check at all, it is a communication check, but it is still one of the key checking points used within a TOC analysis. The whole process of checking the logic of our understanding will be discussed later. Returning to our problem statement, it requires far greater examination. Just what type of communication are we describing? What method of communication? What is the context of the communication? This activity of problem statement refinement continues until a clear, single problem statement is arrived at. Once this clear problem statement has been written down it is possible to consider the organisational impact of this one entity. How much does it cost us on its own, and in both financial and non-financial terms? This gives us a clear understanding of each entity and some indication of what the cumulative impact might be.

The next step is to ask more about the context of each entity. Within the TOC lexicon these entities are called *undesirable effects* (UDEs). The primary analytical tool of the TOC is the cloud. For this particular analysis, the cloud used is called the UDE cloud, which was first described by Dr Goldratt at a conference in London back in 1993. It has the structure shown in Fig. 2.4.

The cloud depicts the conflict between the two statements in the **D** and **D'** boxes, the conflict between what I have and what I want. Filling in the boxes takes time although experience can reduce it to a few minutes. Remember this is not an exercise to be treated lightly. This analysis is intended to reveal key core drivers in a problematic environment and we want to be sure of our ground. The solution will be built on the foundation of our problem analysis and bearing in mind the importance of 'sensitive dependence on initial condition', one of the most important statements of chaos theory, making a mistake here could have dramatic consequences.

Therefore for each problem statement, each UDE, build the UDE cloud. Once the cloud has been constructed it can then be checked, by first reading the cloud, followed by the surfacing of assumptions. The level of rigour applied at this stage is vital if the cloud is to reveal the true nature of the conflict that exists. This starts with the nature of **D** and **D'**.

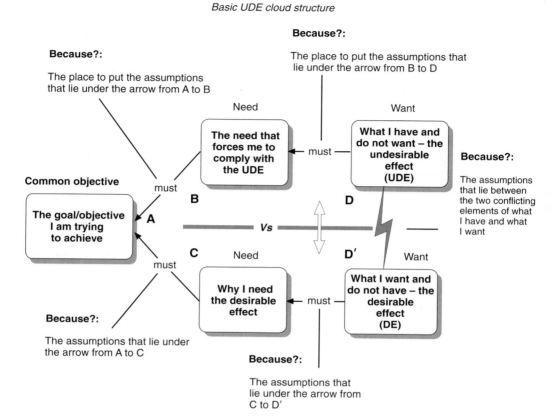

Basic UDE cloud structure

Because?:

The place to put the assumptions that lie under the arrow from B to D

Because?:

The place to put the assumptions that lie under the arrow from A to B

Need

Want

The need that forces me to comply with the UDE

B

← must —

What I have and do not want – the undesirable effect (UDE)

D

Because?:

The assumptions that lie between the two conflicting elements of what I have and what I want

Common objective

must

The goal/objective I am trying to achieve

A

━━━━ Vs ━━━━

C

must

Need

D′

Want

Because?:

The assumptions that lie under the arrow from A to C

Why I need the desirable effect

← must —

What I want and do not have – the desirable effect (DE)

Because?:

The assumptions that lie under the arrow from C to D′

Fig. 2.4. The UDE cloud structure

In order to ensure clarity of conflict, the entries in the boxes must be a clear and precise statement of the entities. They should be written in the present tense and be obviously in conflict. If the conflict is only apparent through the surfacing of assumptions between **D** and **D′** then the verbalisation requires further attention. Once all the boxes have been filled the strength of the cloud should be considered.

This is checked by determining the impact of the cross-connection, that of **D** on **C** and that of **D′** on **B**. If **D** significantly, and negatively, impacts **C**, if **D** places **C** at risk, then that cross-connection is a powerful one. If the same applies to **D′** on **B** then the cloud is a particularly strong one. One such cross-connection is sufficient to give a cloud of some power. Where both cross-connections are strong, then the cloud is extremely powerful. The importance of the cross-connection cannot be underestimated. The cloud is a powerful tool for describing conflict. Often the real impact of the continued existence of the conflict is not properly recognised. The cross-connection is rarely observed in reality, but once seen through the cloud it can be a powerful tool for seeking a resolution that is a win for both sides.

Next read the cloud out aloud starting from the objective **A**. It is read using the terminology of the logic of necessity, which is what the cloud structure represents. Therefore, the reading starts with 'in order to have (A), I must have (C); in order to have (C), I must have (D').' Then back to **A** for the top of the cloud. 'In order to have (A), I must have (B). Then, in order to have/comply with (B), I must have/comply with (D).' It is usually necessary to change the wording to make it flow better and to reflect system authority realities which is in essence a clarity reservation. Once this has been done the cloud is ready for additional work. For further descriptions of the cloud process see Cox and Spencer (1998) and McMullen (1998), Scheinkopf (1999) and Dettmer (1997).

Here is an example from the world of projects. There are three very common UDEs that we come across time and time again. The first UDE is 'often our projects fails to finish on time'; the second UDE is 'often our projects have difficulty staying within budget' and the third is 'often specification/scope is cut from the project'. If we were to build the UDE cloud for each one, and surface the assumptions, they might come out as shown in Fig. 2.5.

This cloud captures the nature of the conflict between on-time finish and being late. The objective is clear, be successful in managing all the projects within the organisation. In order to achieve that we must satisfy each customer and to achieve that we must finish on time, every time. However current reality is that we have to live with the UDE, and the reason is that we comply with the current rules of project management. In other words, in order to be successful we work to the current rules etc. because there is pressure for us to perform right now, because the market will not wait for us, because this is how we are supposed to do projects, and so on. What of the cross-connection? The impact of being late obviously compromises the necessary condition of customer satisfaction. The impact of finishing on time does not appear to have negative impact on the current rules of project management, unless in trying to meet the timescale actions are taken that the normal practices would frown upon. This is precisely what our research has found. That in order to meet the delivery due date, actions are taken that are in direct conflict with the normal rules of project management, the schedules will be changed, and often, resources will be moved around out of sequence, and many more, normal practice may protect the completion date at the expense of scope or budget. The difficulty in this cross check is that normal practice includes at least three separate elements. This is one reason why the cross-connection check is so powerful. All of these issues are usually driven by local management activity.

33

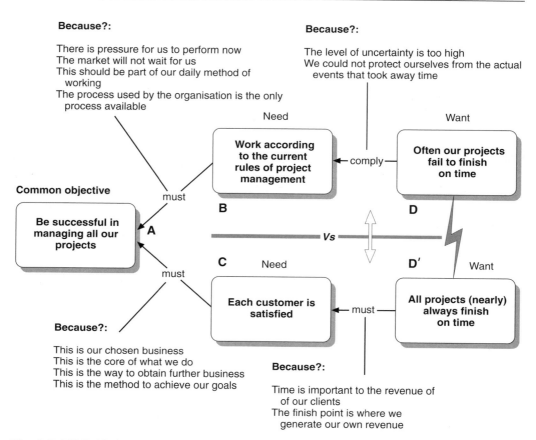

Because?:

There is pressure for us to perform now
The market will not wait for us
This should be part of our daily method of
working
The process used by the organisation is the only
process available

Because?:

The level of uncertainty is too high
We could not protect ourselves from the actual
events that took away time

Need

**Work according
to the current
rules of project
management**

Want

**Often our projects
fail to finish
on time**

←comply—

Common objective

must

B

**Be successful in
managing all our
projects**

A

D

════ Vs ════

C   Need

D'   Want

must

**Each customer is
satisfied**

← must —

**All projects (nearly)
always finish
on time**

Because?:

This is our chosen business
This is the core of what we do
This is the way to obtain further business
This is the method to achieve our goals

Because?:

Time is important to the revenue of
of our clients
The finish point is where we
generate our own revenue

Fig. 2.5. UDE Cloud 1

The second UDE cloud looks like Fig. 2.6.

This cloud focuses on the aspect of budgets and the assumptions are shown. Once more the cross-connection is powerful in both directions. Going over budget either has a negative effect on the external customer, usually because we force him to pay for our inability to work to the original budget, or even though the external client does not see the impact, internal clients do.

The final cloud of the three can be seen in Fig. 2.7.

This completes the three UDE clouds and their respective assumptions.

The first focused on completion date, the second on budget and the third on scope. One interesting aspect of the **B** box of the UDE cloud is that it in itself always contains a hidden conflict. The UDE cloud is part of the organisational construct. The logic of the UDE cloud is primarily driven by the needs of the organisation and the causality that exists within it. However, the continued existence of the UDE cloud is also part of the individual construct and the inherent conflict is between the

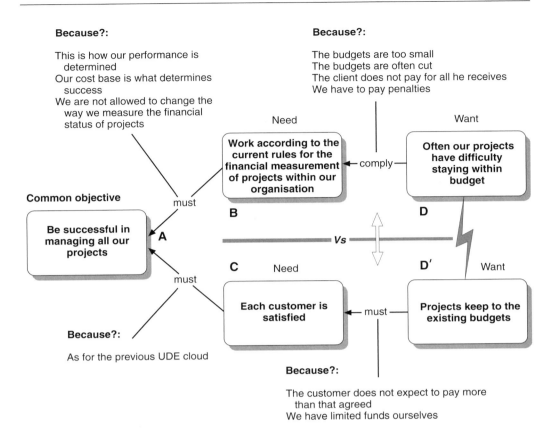

Because?:

This is how our performance is determined
Our cost base is what determines success
We are not allowed to change the way we measure the financial status of projects

Because?:

The budgets are too small
The budgets are often cut
The client does not pay for all he receives
We have to pay penalties

Need

Want

Common objective

Be successful in managing all our projects | A

must

B Work according to the current rules for the financial measurement of projects within our organisation ← comply

D Often our projects have difficulty staying within budget

A

Vs

must

C Need

D' Want

Each customer is satisfied ← must — **Projects keep to the existing budgets**

Because?:

As for the previous UDE cloud

Because?:

The customer does not expect to pay more than that agreed
We have limited funds ourselves

Fig. 2.6. UDE Cloud 2

individual's constructs and the organisational constructs. This is the true nature of the **B** box. It contains a high level choice cloud, a choice between what the individual feels he/she has to make in personal terms set against the need for compliance with organisational requirements. The **B** box highlights the likelihood of a policy constraint in operation.

Therefore, the UDE is both an organisational construct and part of the overall strategy already in place within the organisation. The **B** box is where the individual collides with the strategy, or the policies that support it, head on. The whole of the management of change is derived from this box and the compliance it describes. Once the conflict of the **B** box is recognised, the individual has then to follow the hierarchy of clouds from choice cloud, through decision clouds and conflict of subordination clouds to the ultimate challenge, the paradigm lock cloud. This will be discussed in more detail in the chapters that refer to the change issues within project management.

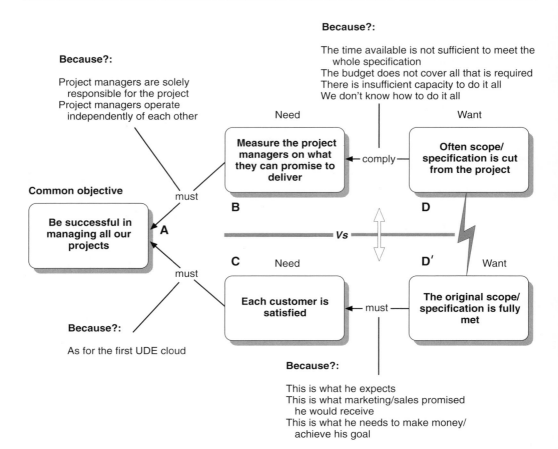

Fig. 2.7. UDE Cloud 3

Having built each cloud, personally checked each one carefully, now check it with a colleague, ensuring that the rules of construction have been followed properly and then check it again. Applying a high degree of rigour at this point pays handsome dividends later. Now comes the really interesting bit, why does this conflict exist? Why don't I just implement the obvious solution? What holds me back? The answer lies hidden in the cloud structure. Each arrow in the cloud contains unwritten statements, statements as yet not revealed. Within the TOC process they are called assumptions.

Assumptions can take many forms. They can be real, true, valid, desirable and positive. They can be untrue, erroneous, undesirable and negative. In fact assumptions can adopt many different guises. The important task now is to bring them to the surface, to reveal the assumptions that hold our conflict in place. Give them prominence, visibility and from that, we gain the ability to challenge them, to find the one, or more than one that is already invalid or can be made so. If we find such assumptions, such erroneous assumptions, we have discovered a gateway to the future. The assumptions that fall into this category allow us to break the hold the UDE cloud has over us once and for all. Remember we are looking for a solution that does not just paper over the cracks, some short-term expediency, but a proper long-term solution that is a win–win and makes a substantial contribution to the achievement of the goal; hence the importance of assumptions.

Now for any one UDE we might stop there. Develop some ideas, which address the erroneous assumptions into fully fledged solutions and go and implement them. This is of course a perfectly valid way to operate, but it is sub-optimal. We could do better.

Consider this, if we have six or seven problem statements, UDEs, from the same environment, and we were to do the cloud for each one, it is a fair bet that many of the same assumptions would appear in more than one cloud. Some assumptions might appear in all the clouds, and some of these assumptions might fall into the erroneous category. That would give us the opportunity to deal with a number of UDEs at the same time. Now the gateway appears to offer a unique opportunity to address more than one issue, and to deliver a substantially more dramatic improvement in our business performance, the key performance indicators first discussed in chapter 1. The process of gaining this level of insight, this potential level of focus and leverage comes from first examining three or four of our clouds and building a composite cloud (Fig. 2.8) that captures fully each of the individual clouds. I have shown only two relationships but the composite cloud needs all such connections to be made. Our example from the three UDE clouds of earlier would then result in the consolidated cloud shown in Fig. 2.9.

The message of this cloud is simple. In order to be successful in the management of projects we must satisfy each and every customer. In order to satisfy each and every customer we must deliver the key performance indicators of price, scope and timely delivery. However, in order to be successful we must comply with the current strategies and measurements used by our organisation, and in complying with these strategies and measurements we have projects that miss the key indicators of scope, time and delivery. The assumption that holds these two in conflict is that there is no other way to address the issues of meeting the key market performance indicators while ensuring that the organisation is meeting the key internal indicators. This also assumes that it is not possible to align both internal and external performance indicators. The cross-connection here is just as powerful as for any one cloud, only more so.

Now that we have completed the analysis this far it is possible to return to the original UDEs and examine their position in the revenue chain. The

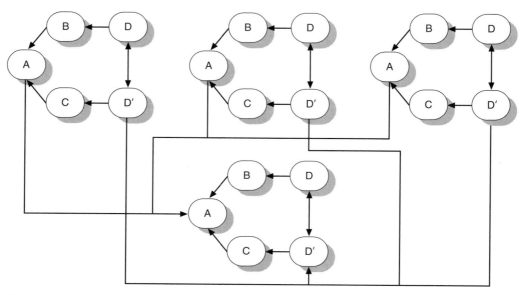

Fig. 2.8. Constructing a composite UDE cloud

UDEs will have come from a number of locations within the revenue chain of the organisation. Using a simple systems mapping process it is possible to describe the revenue chain in more detail and then superimpose the UDEs in their location. It is also possible at this stage to quantify the impact the individual UDEs have in both financial and non-financial

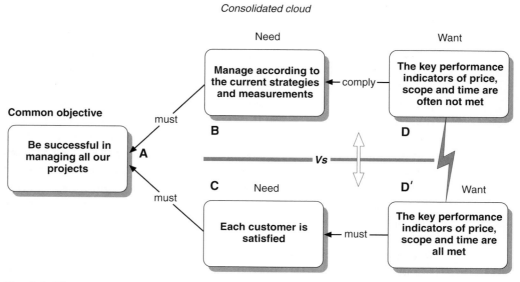

Fig. 2.9. The consolidated cloud

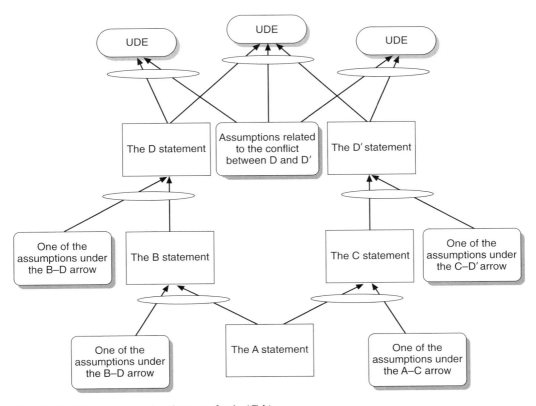

Fig. 2.10. The communication analysis (CA)

terms. Once that has been accomplished the cumulative impact can also be determined.

This analysis, while giving us greater insight into an organisation and some of the problems, is still not good enough for gaining the consensus of all the people involved. We might have constructed an excellent analysis, but as yet we are not ready to communicate this analysis. For that we use another of the logical tools of the TOC, the communication current reality tree (CCRT) or what I prefer to call the *communication analysis* (CA) The simple structure of the CA is shown in Fig. 2.10.

First, it is important to go through an explanation about the construction of the CA. The square boxes are those of the cloud. The rounded boxes are in the most cases assumptions, which in combination with a cloud box give rise to the next box reading from the bottom to the top. There are also times when they are simple statements of current reality. This is the logic of sufficiency and the ellipse represents the 'logical and' link. Reading the CA also follows a different method to that of the cloud, this is the area of the logic of sufficiency rather than necessity, that of the 'if...then' statement.

Thus if the CA above is read out aloud it sounds like this: '**if** the A box **and if** the assumption under the arrow A–B **then** the B box', and so on until the whole analysis has been read. Once more the sense and logic of what we have written has to be validated. The primary tests to be used at this point are as follows

- Clarity: although clearly a logical reservation this is more easily used as a communication or sanity check. Is what I have just said clear? Do I have to explain what I have just said in order for people to see what I am getting at? Does it make sense?
- Entity: is the entity clear? Is it precise? It is a single entity? Does it have causality written into it? Does it exist, or is it perceived to exist now?
- Causality: do I believe the arrow? This reservation has no problems with the entities at either end of the arrow, it centres on the arrow itself, do I believe that the one entity causes the other?
- Insufficiency: this refers to the need for a 'logical and' statement to support the single line causality already drawn. In other words when the logical connection of 'if A then B' is considered to be insufficient, it should rewritten as 'if A **and** if C then B'; in other words A on its own is insufficient to cause B.

The CA has to be subjected to scrutiny using these tools if we are to have confidence in the analysis and use it as the foundation for developing our solution. Equally, if the CA is not clear we will fail to gain the consensus we seek.

The CA can be done at any stage, using any UDE cloud. In each of the case studies described in the next chapter the three functional heads did it for their own specific areas as well as the enterprise wide analysis. I have used it on single UDE clouds in order to address specific operational problems, one such use being how to deal with specific buffer violations within critical chain project management, of which more later. But first here is the CA from our example (Fig. 2.11).

This CA shows the real impact of the conflict. We start at the bottom of the page and work up. So, if we are to read the CA properly it starts with

*If I want to be successful in managing all our projects, **and if** a necessary condition for the future growth of the company is customer satisfaction, **then** we must have each customer satisfied (with what we have accomplished for them). **If** we must have each customer satisfied (with what we have accomplished for them) **and if** the key performance indicators are fundamental to customer satisfaction, **then** we are under pressure to meet the key performance indicators of price, scope and time. However, **If** I want to be successful in managing all our projects **and if** corporate measures determine the*

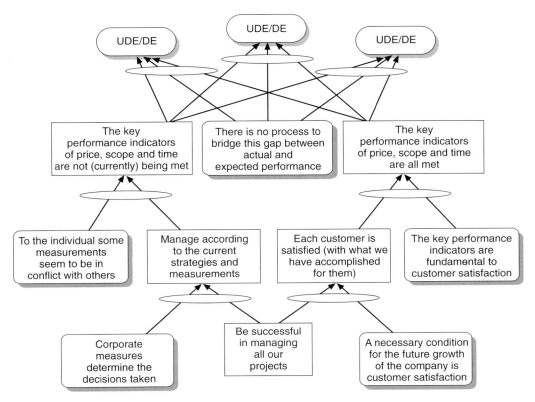

Fig. 2.11. The CA of our project example

> *decisions taken,* **then** *we manage according to the current strategies and measurements.* **If** *we manage according to the current strategies and measurements,* **and if** *to the individual some measurements seem to be in conflict with others,* **then** *the key performance indicators of price, scope and time are (currently) being met.* **If** *there is pressure to deliver according to the key performance indicators,* **and if** *the reality is that they are not being met,* **and if** *there is no process to bridge the gap between actual and expected performance,* **then** *other UDEs will occur.*

I am sure that you can suggest some if the scenario appears familiar! From this analysis the areas to address might be, first, addressing the apparent conflict between the measurements, and the second is to invalidate the assumption that there is no process, which can bridge the gap in performance. How this is done is later in the analytical process of the TOC.

What we have achieved so far is a process of analysis that stretches from vague, fuzzy problems we are currently having to try to manage to a rigorous, logical statement of the core drivers for the area of the organisation under our control, and if we are the top team, then we have an enterprise wide analysis of our revenue chain.

We are now in possession of information upon which to consider some strategic decisions. The primary drivers, good and bad, functional and dysfunctional, contributing and detracting from performance, have been identified and their causality recognised in a formal, structured manner. The gateway to addressing the key business issues now lies open before us. We have the means of communicating this analysis and then gaining the buy-in of not just the key people, but all the players in the revenue chain. This includes subcontractors, key suppliers, clients, outside agencies and so on. They can now all see the nature of the revenue chain and the key drivers preventing all from gaining substantial benefit.

Having such a clear statement of the problem area and the core drivers is only the first step in the TOC change model. At this point, given the ability to properly communicate the analysis, it is possible to have gained consensus on the problem. The second step is to build a clear statement of the direction of the solution. The starting point for this is called, within the TOC lexicon, an *injection*. An injection is defined as 'a statement of actions completed'. It is written in the present tense and contains a single statement. A better description is that the solution that successfully deals with the cloud contains at least one feature, the injection. It is likely that more than one feature will be necessary to successfully overcome the problems contained within the cloud analysis. Indeed in all TOC applications the solution contains a number of features, which must all be implemented properly for the solution to be truly effective.

Two examples of such features from the implementation process of critical chain are, first, 'project managers have the project's network of tasks and path dependencies – identified by resources'. And secondly, 'the resources work in a way that minimises the wasting of the task duration times that have much less safety time contained within them'. Note both are in the present tense, even though they do not yet exist. They also contain a significant number of actions. We might not even know what actions are going to be necessary to achieve the feature. There is therefore a great deal of work still to be done. The primary reason for writing them in the present tense is that when we wish to validate the solution, the assumption is that every feature has been completed; the logic is 'if I have all the features **then** I have all the desirable effects'. This allows us to check the validity and robustness of our solution and to verify that it will deliver what we need in order to address both the constraint and achieve our goal. The actual process of developing the plan to achieve the feature is done at a later stage of the overall enterprise analysis.

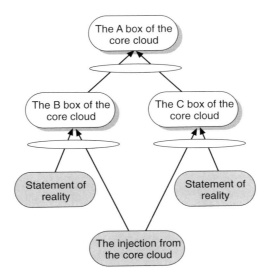

Fig. 2.12. The Core FRT

From our worked example the two starting features might be, 'all measurements and strategies are aligned' and the second might be 'the process to bridge the performance gap is implemented'. The assumption at this stage being that if we have both of these features it is possible to achieve substantial improvement towards the goal of successfully managing all our projects.

Therefore stating the feature is not enough if we are to have confidence that the direction of the solution is right, we have to connect the feature(s) to at least those positive aspects of our clouds, the DEs described in each of the **D′** boxes. This applies to the individual UDE clouds, the composite clouds and the enterprise cloud. Even better we should also connect to the other DEs that exist bearing in mind that we did not use all the UDEs at the start, and for each UDE there is a corresponding DE. This not only shows the enterprise wide nature of the solution, but also gives each UDE owner a real sense of involvement and the realisation that the enterprise wide solution will have real beneficial impact at the personal level.

The TOC construction that presents this analysis is called the *core future reality tree* (CFRT). The structure looks like Fig. 2.12.

When checking the analysis it might be necessary to add further features/injections to the original in order to secure the solution. This might mean the logical structure ends up looking more like Fig. 2.13.

If we take the worked example then the construction of the core FRT might look something like Fig. 2.14.

Note that there are entities that also existed in the CA represented once more. That is simply because they were valid then and are still valid today,

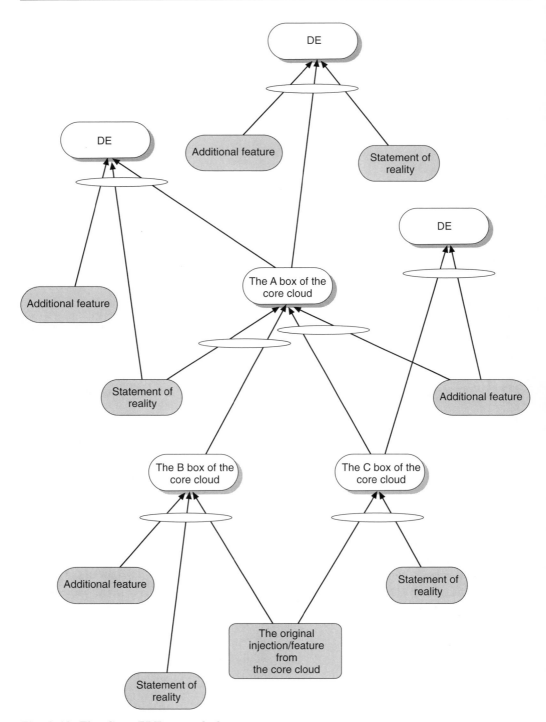

Fig. 2.13. The Core FRT extended

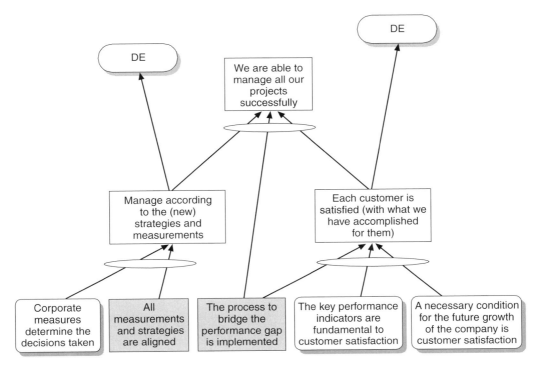

Fig. 2.14. The Core FRT of our projects

i.e. in the future. The DEs are derived from the UDE clouds and also from other UDEs not used in this initial analysis. Once this has been done it is possible to communicate the solution and gain the consensus of the people to the direction of the solution.

The third stage is gaining consensus on the benefits of the solution. This involves checking the impact of achieving each and every DE in both financial and non-financial terms. This gives real clarity to the organisational pay-off in implementing the solution. It also works at the individual and departmental level. It should also give real clarity, focus and leverage to the revenue chain of the enterprise. As with the initial analysis this is not an area for superficial analysis and rigour. The solution must deliver the DEs and therefore the robustness of the construction and the validity of the logic are paramount. Once more the logical reservations are used to determine our confidence level in the solution.

The next stage is to develop the implementation plan. What is it that we have to do in order to implement all the features the solution contains, in other words how do we get to where we want to be?

We have defined a feature as a statement of actions completed, but what actions? Equally the development of the implementation plan gives rise to more than one type of reservation. There is perhaps the most obvious

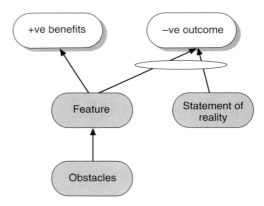

Fig. 2.15. Dealing with reservations

which is 'I can see the objectives but there is plenty to be overcome if we are to succeed'. This is the identification of the barriers standing in our path, the barriers to the implementation, what the TOC process calls *obstacles*. The source of these obstacles is easy, all the people involved in the implementation process will have obstacles. They will all be able to suggest candidates for the obstacle list. They come from experience, past failures, being ignored in the past, and having their experience belittled. These reservations all share one key aspect, they have to be overcome in order for the plan to be implemented, the feature to be achieved.

The second type of reservation comes as a result of considering the combination of having implemented the feature in conjunction with something, which already exists within the organisation and its current environment. This is shown in Fig. 2.15.

The tool within the TOC that deals with obstacles and develops the action plan is called the prerequisite tree (PRT). The tool within the TOC that deals with the potential negative outcomes is called the negative branch reservation (NBR). By encouraging the people to raise both types of reservation it is possible to achieve the fourth stage of successful change – dealing successfully with the reservations of the people. Once the PRT has been built, and the NBRs incorporated with the original plan there is only one thing left to do – make it happen. This is the area of the transition tree, perhaps the most powerful tool of the TOC/TP after the cloud.

A typology of clouds

The research that has been carried out in the application of the TOC/TP has led to the creation of a typology of clouds. There is in fact both a typology and a hierarchy of clouds. The hierarchy was first introduced

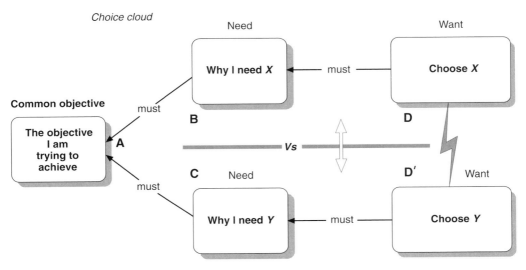

Fig. 2.16. The basic choice cloud structure

briefly in chapter 1 and will be reviewed here. All clouds are set into the context of choice. The initial decision is about choice. At its simplest level the choice cloud would look like Fig. 2.16.

I cannot choose both *X* and *Y*, I have to make a choice. The statement in the **B** box is the need that the choice of *X* supports, equally the statement in the **C** box is the need that the choice of *Y* supports. Note that both **B** and **C** are necessary conditions for the achievement of **A**, hence my problem. The following choice cloud in Fig. 2.17 is a common example in many organisations.

Note that I have not included the needs. The key element of a choice cloud is that they are rarely broken, people simply choose which side they are going to work on and take it from there. There is no attempt to try to understand the cloud in more detail, only choice. There could be many entities that represent quite acceptable needs. The important dimension is that having made the choice, all further decisions and outcomes are driven by that choice.

The next cloud is drawn from the choice having been made. This is the decision cloud. They are still very much part of the choice cloud, and indeed may look the same. The difference is that the choice cloud is a conflict between two philosophies, two generic approaches, the decision cloud is between the hard practical implications from both potential choices that the choice cloud encapsulates. Given the choice cloud above, the decision cloud in Fig. 2.18 is typical.

There are more decision clouds driven by the same choice cloud, this is but one example. This is where the hard decisions have to be made and

47

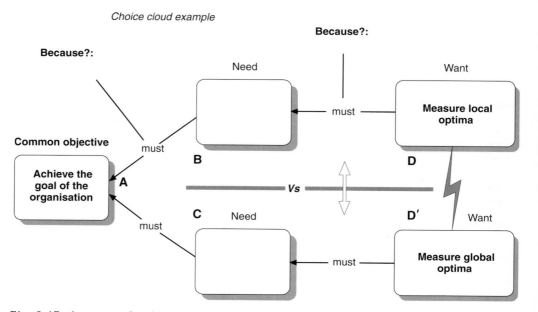

Fig. 2.17. An example of a choice cloud

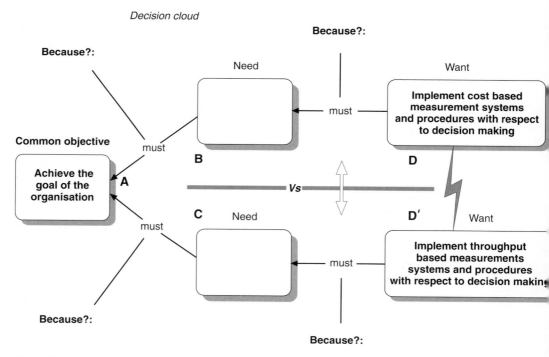

Fig. 2.18. A decision cloud

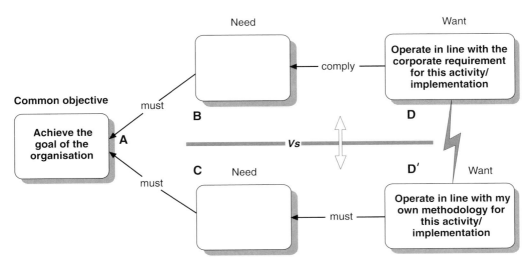

Fig. 2.19. Conflict of subordination cloud

the process of surfacing assumptions becomes the most important activity. But there are times when even though the decision has been made, an individual feels that the decision is not in line with his/her intuition and has great difficulty complying with the decision. This leads to the 'conflict of subordination' cloud as shown in Fig. 2.19.

This cloud can still be overcome through the usual process of surfacing assumptions and developing a workable solution that satisfies the requirement of win–win. However there are still occasions when the individual cannot under any circumstances accept the decision that has been taken. The very idea of complying with the decision is anathema to the individual, it is a threat to their sense of being, it is a challenge to their paradigm about the issues being discussed. This leads to the final cloud in this hierarchy, the paradigm lock cloud as shown in Fig. 2.20.

This cloud was the subject of my research at Cranfield University and is the basis for the major dysfunctional obstacle in the management of change. It will be discussed in more detail when we come to the implementation strategies surrounding enterprise project management. The cloud hierarchy described here was a key finding of the research into TOC implementations. Many times it was found that this hierarchy determined behaviour patterns within the organisation. Each of these clouds represents the individual conflicts with which people have to deal.

The discussion so far has focused on the clouds that are part of the individual. They are part of the construct each individual has. In addition

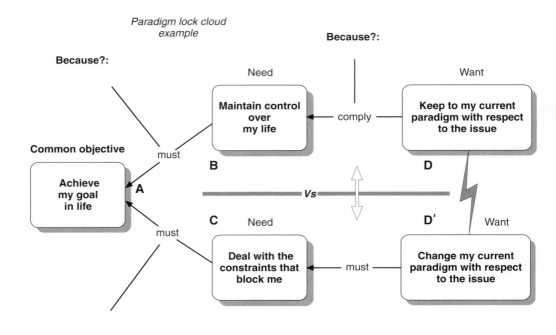

Fig. 2.20. *The paradigm lock cloud*

there are a set of organisational clouds that must be addressed. We have already discussed the UDE cloud. The next type of cloud is that of the conflict cloud. Here the conflict is between two people and can be seen in Fig. 2.21.

Other clouds that are used within the TOC process are the core problem clouds, the obstacle clouds and the subordinate's cloud. These clouds and how to use them can be found in the work of Scheinkopf (1999). The full process used within the TOC enterprise analysis, and which was used to capture the case studies introduced in the next chapter is depicted in Fig. 2.22.

This process was used in the following case studies in the analysis of the project environments that formed the basis for both the research and this book. Although more comprehensive than what has been discussed so far the basic outline follows the same path. The intention is to end up with an analysis that can be used for both communication and confirmation of the problem area and the core drivers that are in operation. Once this has been achieved the next stage is to develop the solution and that is shown in Fig. 2.23.

At this point it is possible to then implement the whole solution, obtain the expected results and move on to the next problem, using the same process once more, and continuing to do so until the highest level of

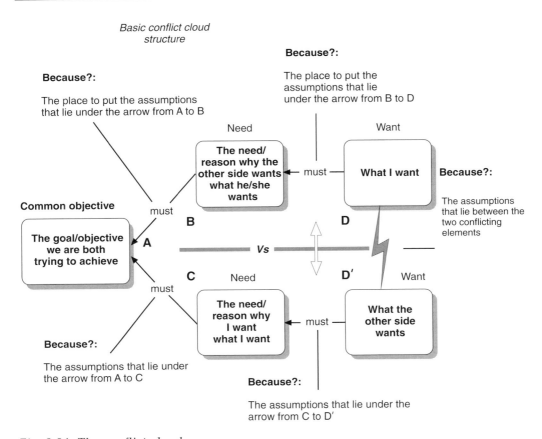

Fig. 2.21. The conflict cloud

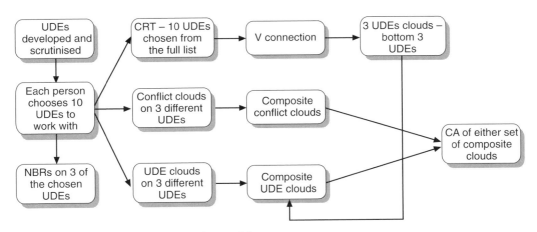

Fig. 2.22. The development of the problem

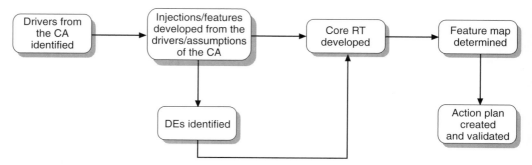

Fig. 2.23. Developing the solution

performance has been achieved. This whole process however is conditioned by one very important set of what I have called the *laws of
constraint management*. Dr Goldratt founded the whole movement of
the Theory of Constraints. It was his imagination and foresight coupled
with an enormous ability to think in a new and challenging fashion that
led to the creation of the TOC and the thinking processes. It was through
his ability to sit with others and develop the communication tools that the
TOC is now a generic problem-solving tool available throughout the world
and used by many different companies and organisations. What I have
found in my work with TOC is that every implementation is bounded by
the laws which surround the constraint. The final element of this chapter,
before moving to the primary case studies and the issues that arose out of
them, is to consider what these laws are and allow for a discussion about
their impact later in the book.

The laws of constraint management

While involved in many TOC/TP activities during the last ten years, I have
noticed that in every activity what could be described as a series of truths, a
series of what I now call laws, exist. They might appear to some to be statements of the obvious, and maybe they are. But equally they are ignored at
some peril and therefore are worthy of recognition. They should also be
thought of in the context of some of the business performance indicators
discussed in the first chapter. These include shareholder value, profitability,
productivity, time to market, resource effectiveness, sound financial reporting systems, sound decision-making processes and so on. I consider there to
be seven such laws of constraint management.

- The first law is that every organisation is a system comprising interdependent elements that form a revenue chain.
- The second law is that the ability of the revenue chain to maximise
  performance is determined by the weakest link – the constraint.

- The third law is that the constraint is the primary location for both focus and leverage for the improvement of the overall performance of the system.
- The fourth law is that improving any other link in the chain does not improve the overall performance of the chain itself.
- The fifth law is that subordination to the constraint in terms of the measurement system, the policies of the organisation, and the way in which the people operate, is a fundamental requirement of managing the chain.
- The sixth law is that the management of the constraint, and non-constraints, is dependent upon the use of an effective decision support system.
- The seventh law is that variation in the system has most impact on the constraint.

The variation we are thinking about here is special cause variation rather than common cause. The last law also argues that variation at a non-constraint is therefore far less important. This leads to the necessary conditions of both a sound management methodology and a decision support system, which ensures that actions, as a result of such variation, are undertaken only when the management system says so. This is primarily when either the goal, or at least one necessary condition to achieve the goal, is being violated.

Having the laws is insufficient. There has to be a process of applying them to an organisation. This process has to be robust, it has to be rigorous, it has to be logical, it has to address dysfunctional elements within the organisation, it has to be capable of enterprise wide analysis. The process used within this book is the TOC/TP. Having a process without recognising the impact of the laws is just as (bad) as having the laws without a process. Now if we combine the process described earlier and the laws as discussed here, it is possible to address the core business issues and deliver improvement in the key performance indicators described in chapter 1.

CHAPTER THREE

The introduction of the TOC solution – critical chain project management

Introduction

This chapter introduces the solution developed within the TOC for project management entitled *critical chain* and how it is used within project environments. It is not the intention to repeat here what has been written so well by authors such as Larry Leach and Rob Newbold. The technical detail of critical chain has been covered in great detail by them and needs no great introduction here. This chapter is about giving sufficient under-standing of the basic concepts of critical chain such that the reader can see what it is trying to achieve, and set the approach into the context of an enterprise focus for projects and project management in general. The approach came about because people who saw the elegance and simplicity of the TOC solution for the manufacturing environment and production scheduling, saw that it could be transferred to the world of projects. The same importance attached to the notion of the constraint; the same impor-tance of safety and the right place to put it, these and other aspects of the production solution were transported to project management.

Background to projects and project management

Every project is supposed to have an objective. Some are fairly obvious such as the completion of a road, bypass, sewage works and so on. If the intention is to build a new airport, or a new stadium to hold a major championships then whether the project is completed on time or not is clear for all to see. But projects often have more than one objective. The companies involved expect to make money out of the project. The client intends to gain revenue from the product, the contractor expects to make money from his labour and so on. The downside of non-compliance with the demand of the project is that penalties may well be applied, and these financial penalties come out of the profit. Equally, any work not completed by the due date will still have to be completed, and in many Design, Build and Finance projects the remedial work is carried out by the contractor at his expense. This further drains the profit pot.

Within the high-tech industry the intention of the project is to bring to market as early as possible products that substantially improve the bottom line. The penalty here for being late is that someone else will take the market, often completely. One important distinction between the two sectors that became apparent during the research involved profit margin. With the construction companies many of the projects examined had a small margin for profit, often between 2–3%. Even in some of the larger projects the margin was still relatively small, between 8–9%. In the high-tech market the picture was very different. The scale of the projects in budgetary terms was often higher, and the corresponding margin also higher, between 25–30%. This difference had a profound affect on investment decisions. The paradigm within most civil engineering companies is that innovation projects, such as critical chain, are financed by the individual project, rather than the centre. With such small margins this does not leave a great deal of flexibility for real investment. Within high-tech the environment allows for investment in such innovation either spread across a number of projects, or funded through the centre thus lifting the pressure from any one project, or group of projects.

Typology of projects

But are all projects the same? Within the work undertaken over the last five years three distinct types of projects have been developed. These are single project, multiple single project and multi-project environments.

Single project environments are perhaps the simplest to work in. They are easily defined within an organisation as stand-alone projects. The people working on them only work on that one project. The equipment is allocated to that single project and remains with the project until completion, or at least the completion of their involvement. Single projects can still be quite complex with many activities to be completed both consecutively and concurrently. The scheduling exercises still have to be done. The work breakdown structure still requires to be completed. Indeed, all of the usual activities of project management apply. Risk analysis, cost breakdown, procurement and so on are part of the dependent events that form the chain of activities leading to a successful conclusion.

Multiple single projects is simply the environment where the organisation is structured to deliver more than one project at the same time but has no, or a very limited, overlap of resources. Each project is still stand alone with resources usually being allocated only to that project and to no other. Many construction companies are structured this way. Where there are resources allocated across projects they are never key resources, never scarce resources, usually falling into the category of HQ staff people, support people and so on. For each single project all of the usual activities described above still apply.

Multi-project environments are where there is considerable overlap between projects in terms of resources. This is very much the scenario for most high-tech companies. Resources are allocated to a number of current projects, perhaps even still working on projects that require some form of remedial activity, and possibly preparation for new projects as well. Even within a single project they will be working on a number of concurrent activities. The pressure here is usually immense. Clients, and their representatives, are continually seeking earlier and earlier delivery dates. The management seek to launch new projects before existing projects have been cleared out of the project department. The ability to prioritise within these environments is usually based on a basis of who shouted the loudest last! At times there are many people all shouting at the same time and at the same volume level. Success in a multi-project environment however is incredibly lucrative. If a set of projects can hit the market early then the financial implications far outweigh any local expenditure constraint. It is into these types of projects that the critical chain approach was developed, and implemented. It has worked in all three environments.

Critical chain introduced

The critical chain approach to project management was first introduced by Dr Goldratt in his book entitled *Critical chain* in 1997. However, the approach had been used within the TOC environment for some eight or nine years prior to the book being published. Subsequently others, notably Newbold (1998) and Leach (2000), have written excellent books on the subject. I do not intend here to reproduce their work, but I think it necessary to give some overview of the approach and if the reader wishes to understand more then the work of these other two, in particular Leach is highly recommended.

Creating a network of activities to form a project plan is the starting point of network creation. This, of course, assumes that the goal of the project has already been determined. Most projects have more than one activity. That being the case it is important to build the network of activities keeping the logical structure intact. By this I mean the events that are dependent on each other should be reflected in the plan. Once all the activities and the dependencies are known it is possible to construct the plan either using PERT networks or Gantt charts. The starting point is the same in both cases, how long is each activity going to take – in other words an estimate of the length of each activity. Usually the people who are going to do the task are invited to give such estimates. Therefore for any one task the estimate looks like Fig. 3.1.

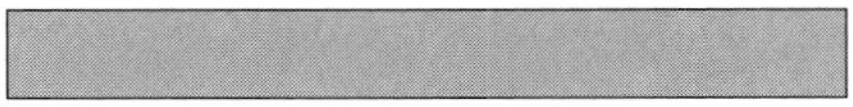

Fig. 3.1. Actual time required

The bar represents a period of time equal to the estimation. But this is usually based on all things going well, but how often do things go wrong? There can be substantial variation during execution. The task might prove to be more difficult than previously thought. The technology might not work. There might be rework necessary before handover, and many, many more. This means that there is a level of uncertainty in the estimation, which has not yet been included. Therefore, in order to accommodate the possibility of such variation the bar changes to that shown in Fig. 3.2.

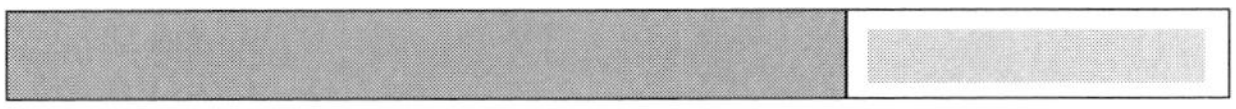

Fig. 3.2. Actual time plus some time for variation

Although the overall time has been increased the potential for variation has been covered. But what if I have to do other things in the same time frame? What about other projects I am involved with? What about the various meetings I have to attend? What about the other responsibilities I have to discharge in the course of a normal day? I had better include all of these in my estimation. This results in a further extension (Fig. 3.3) to the original estimation.

Fig. 3.3. Actual, variation and some time for other activities

But how often in the working day am I interrupted? There are always interruptions taking place, and all of the time it seems. The telephone rings, e-mails to catch up on, the dreaded knock on the door, a whole plethora of small events that somehow add up to most of the working day! Every time it happens I lose time on all the other activities I am supposed to complete, therefore I had better include an element for that as well (Fig. 3.4).

Fig. 3.4. Actual, variation, other activities and some time for interruptions

Now I have an estimation to be proud of! I have included the time to complete the task itself, I have included time for the possibility of variation within that task, I have included time for all the other tasks I am responsible for at the same time, and finally I have covered myself from the variation that occurs as a result of interruptions. This leaves the estimation looking like this Fig. 3.5.

Fig. 3.5. The estimation given!

The next question is related to the time we are prepared to commit to. Once we move to execution it is likely that most times will finish as shown in Fig. 3.6.

Note that the statistics show that we hit the actual time we said, in other words we appear to finish early compared to the estimation given (elapsed). It should be noted that there are times when the activity did not finish around the actual estimation but way beyond it. In other words we needed some, and at times all, of the protection we felt was necessary. There will even be times when the time required was even further to the right than t_4 in Fig. 3.5. In the cases where we were clearly late there are of course many reasons that we can quote as to why. The result is the same however, we were late and that conditions the next estimation.

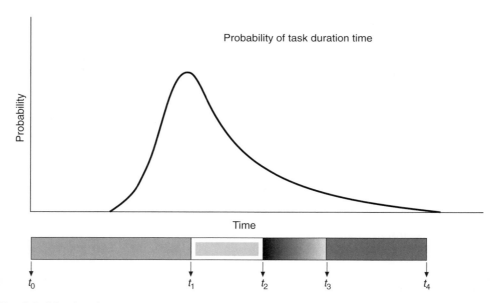

Fig. 3.6. Moving from estimation to execution

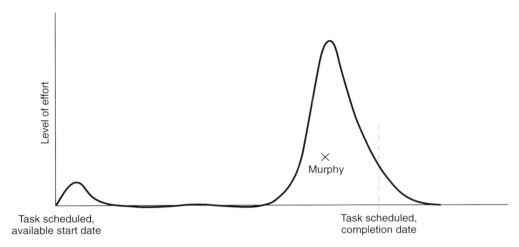

Fig. 3.7. What happens when we delay starting the activity?

Giving a commitment in the terms of an estimation is a different exercise to where we think the activity might finish. Experience has taught us that if we give a commitment and fail to meet that commitment there is the likelihood of being taken to task about our non-performance. It is always preferable to err on the side of caution. Sooner be applauded for completion on time, with a reasonably large estimation, than be penalised for finishing late with a small estimation. In many companies, in many project environments, the usual reaction to such events is to work with the largest estimation allowed and then work carefully to it. If manage-ment cut the estimation there is still sufficient protection to allow me some degree of comfort.

There is a danger however with such estimations, I call them elapsed estimations, rather than the actual estimations. We feel that there is plenty of time, so often the start of the activity is delayed. We complete other tasks, we allow interruptions, we attend other activities and so on, without necessarily starting the task we were supposed to, at the time we thought we might. This leads to the so-called 'student syndrome'; putting off the time to start, knowing that there is plenty of time to complete. At least according to the elapsed estimation there is. The reality is different however as can be seen in Fig. 3.7.

Here the task has been delayed for other events to take place. I now start the activity with much less time left within the estimation to give the level of protection I felt was essential when first the estimation was given. But I have done nothing to reduce the possibility of further interruptions, or variation within the task itself.

Now the task starts later, and I have done nothing to change the content of the task so the possibility of variation has not changed (Fig. 3.8). There

59

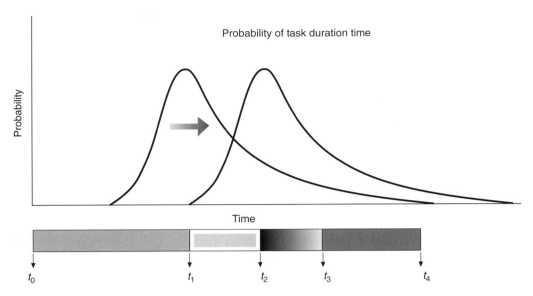

Fig. 3.8. The real pressure on the project

is every chance that the completion date will be put in jeopardy. Although the actual time to complete may not differ much, the time at which the activity is completed has been shifted along the x axis by a considerable margin and the likelihood of missing the due date is now very high. Equally, all subsequent tasks also run the risk of starting late as a result.

In the construction of the critical chain schedule the key aspect is to work with the original estimations, the actual estimations rather then the elapsed estimations and account for the possibility of variation at a different location, and work the project differently to remove the possibility of negative impact from other project activities and interruptions.

A description of critical chain starts with the recognition that the constraint of any project is the longest chain of dependent events by resource. This is what dictates the overall lead-time of the project and is therefore *de facto* the constraint. Remember the constraint is defined as anything that prevents the system from achieving the goal set for it. The goal of a project is to deliver a product or service by a certain time, the due date. The faster the due date can be achieved the faster the system achieves the goal. This is why time, the time from the start of the project to full completion, with no degradation of specification, and full compliance with the budget, is the constraint of any project.

This is, of course, dependent on the ability of the team to produce the network of activities properly and completely for the lifetime of the project. It also means that the management have the necessary tools to identify the critical chain. Certainly within our experience we have identified the critical chain without the need for specialist software, although having that facility does help. Within multi-project environments however the use of appropriate software is essential. However, assuming we have the necessary tools, at this point we should have the constraint of the project clearly identified.

A worked example – single projects

Consider the following project network. This is a simple project and has been produced using the rules and procedures of critical chain. Estimations have been used that reflect the actual time to complete each task rather than the elapsed time. The logical structure has been validated, in other words the activities have been checked in terms of both preceding and successive activities. Each colour code represents a single resource, each different from any other. The network has been approved by the key project people; the project leader, the team leaders and the client. Remember as the estimated times are actual and not elapsed there is little if any protective time, or float, in any of them.

Although the overall time to completion may seem short, there is currently no protection for variation of either special cause or common cause. Also note that the project activities have no start and finish times, only durations. Neither are there any milestones. Often these fixed times act as an additional constraint and/measurement for the project. However, TOC argues that the success of the project is not determined by any internal milestone or dates fixed in the calendar by the software, but by the overall completion of the project. Hence the importance of task durations during the estimation phase.

The next step is to identify the critical chain, the constraint of the project. This is defined as the longest path of dependent events, by resource. As a matter of interest the critical path (Fig. 3.9) is from M1 through G2, M3, B4, B5 and Y4 to completion. The critical chain, however, is different and is shown in Fig. 3.10 with activities shaded.

Therefore the critical chain activities are R1, Y2, Gr3, G4, G5 and Y4. All the other activities are feeding that chain. However the time to completion is not the sum of the critical chain activities, there is still the question of protection. The normal protection contained within every activity is not present therefore protection in the form of buffers must be included. This involves both sizing and placement. The first buffer to work on is that concerning the resources. It is vital that the key people are available on or slightly ahead of the time they are due to work on the project. If they are late, then the likelihood of the project running late is high, they are working after all on the critical chain, and therefore on the constraint of

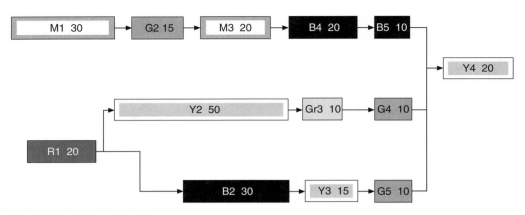

Fig. 3.9. Initial project layout

the project. Any delay on one critical chain activity places a delay onto the whole project.

Resource buffers are used then to give an early warning to the resources, and the resource managers, to allow them time to prepare for the work they have to do. Equally, if the preceding activity finishes early then we want to take advantage of such an early finish and move

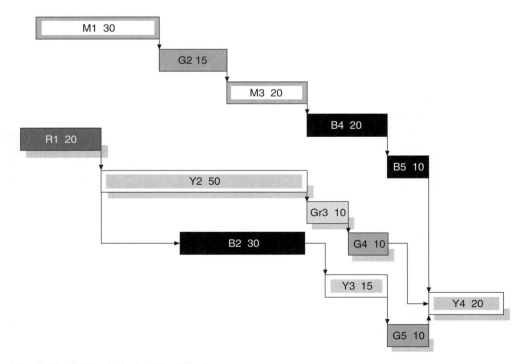

Fig. 3.10. Critical chain identified

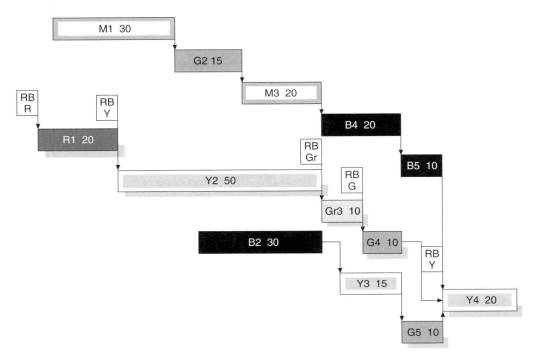

Fig. 3.11. Resource buffers placed

quickly to the next activity. If the resource is waiting for the original start time, or continues to work to a computer generated start-date or policy determined milestone, then we have lost the opportunity to move the project forward, and lost the opportunity to perhaps finish the whole project early. Thus the resource buffer (Fig. 3.11) is an important aspect of the critical chain approach.

One final aspect of the resource buffer concerns the ownership of the resource. There are two key managers in this project, the project manager and the resource manager. The commencement of the resource buffer heralds the shift of control of the resource from the resource manager to the project manager where it remains until the task is fully completed.

Once the resource buffer has been sized and placed, the next step is to protect the project from variation within the tasks themselves. Thus the critical chain and the feeding chains require protection. These buffers are called the project buffer and the feeding buffer respectively. They are situated at the end of the chain they are protecting and sized according to the level of risk within that chain. This level of risk takes note of all the usual sources of paranoia in the minds of the people doing the tasks, and those managing them.

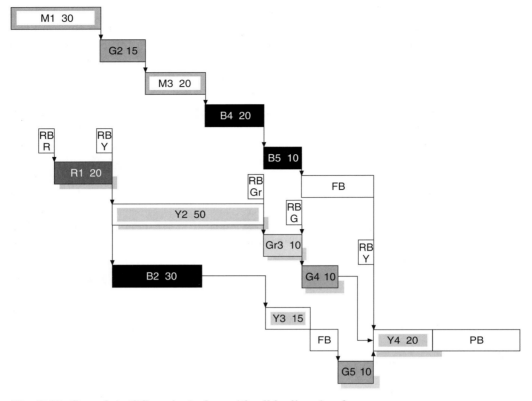

Fig. 3.12. Complete CC project plan with all buffers in place

Now we have the final project structure with the buffers in place (Fig. 3.12). Each resource buffer is in place to protect the start of the critical chain activity for its respective resource, and the feeding buffers and project buffer are also ready to do their job. It is often noted that there are activities that begin prior to the critical chain activities, which leads to the suggestion that the critical chain has moved. The critical chain is defined prior to buffer insertion. Because M1 starts before the first activity of the critical chain R1, it does not make the constraint of the project shift to that arm of the project network. The longest set of dependent activities is still from R1 through to Y4. In order to ensure that the chain from M1 to B5 does not impact the constraint, the feeding buffer is inserted and the size of the buffer reflects the level of paranoia the people have about that set of activities.

What we have now is a project plan with all activities utilising actual times, not elapsed, with the buffers all in place and the due date checked. If the overall time envelope of the project plan, including buffers, exceeds the overall time allotted for the project it is necessary to elevate the constraint. This might mean checking the logic linkages between activities, adding more resources, subcontracting some activities so that they might be able to be completed in parallel with other tasks and so on. At the end of all this activity the project plan can be approved and implemented. We are ready for the transition from estimation to execution.

In execution the key question to ask is 'what is the remaining duration of the current tasks?'. This gives much more information than simply asking how much has been done. Time is the critical factor, not work done and knowing how much time is left gives a clear indication about the robustness of the plan, and the implementation. Also we are seeking early finishes being handed on so there is an expectation that the initial buffer size will increase as activities finish early, and decrease when they finish late. The buffer time is therefore dynamic once we are in execution. Using the zone management approach to the buffers gives us a clear indication as to whether any penetration is due to special or common cause variation. The level of penetration determines the actions necessary to address the scale of the problem. All focus is now on the completion of the project on or ahead of time, not reacting without reason to the delays that might occur to any one task, or set of tasks.

The final element in the implementation is to ensure that all the people involved work according to the new rules of managing projects. This is the dimension of subordination referred to in the five steps of focusing. Once the decision has been taken to implement critical chain then there is no further debate, subordination is the order of the day. Early finishes are passed on correctly, the work being undertaken is done in such a way the problems that might have been common in the past are no longer there. There is little or no multi-tasking, people do not wait to start activities, if an activity has been released to them they start without delay and work through to completion properly. Early finish is not a concern that might result in estimations in the future being reduced. Early finishes just reflect statistical fluctuation and nothing more.

This leads us to consider the behaviours that exist in a critical chain environment. The intention is to have an open and honest culture where people are encouraged to work together as a team, often across functions and levels. There is no blame culture in such an environment. In terms of network planning the actual estimations are freely given, network validation and resource definition are exercises to assist the project to completion, not protecting oneself from attack. Tasks are only carried out in line with the plan and the relay runner ethic dominates. This means that people run as fast as they can and hand over in a clean and effective manner. People are only focusing on the tasks in hand and no other thus reducing the level of multi-tasking, they report duration to completion, they report early finishes and if there are problems they report them as early as possible.

Buffer management is the key to the successful operation of the plan, and the buffer reports go to the key people within the project, and the key line managers within the organisation, and also to those key people outside the immediate project area. This might include the client, the suppliers and always includes the senior management levels. The expected effects on the projects are on-time or early completion, no reduction in specification, good budget control over the lifetime of the project, better safety statistics, reduced firefighting, higher morale and significantly increased margin. In terms of the organisation the expectation is that the company is more competitive, it is able to attract higher market share, it can achieve higher levels of turnover, it can choose which projects to take on board, it can achieve the profit target set and is in the position of ensuring that the client is satisfied and that repeat business becomes the order of the day. If we can do all that then we have achieved the goal set out in chapter 1 and the necessary conditions that must be met.

The sequence of creating a critical chain schedule ready for implementation

Any enterprise resource planning type application such as critical chain, which uses the TOC process of analysis and solution development, will contain a series of features. These features will refer to both software and the management methodology. It is important that all the features contained within the solution, both the initial generic solution and the final specific solution are properly understood, validated and implemented. Therefore in order to deliver the critical chain solution, there are a number of features that must be achieved. These features fall into two distinct categories, the first refer to the generic elements that are common to all critical chain project implementations, the second to those aspects which are specific to the environment in which the solution is being implemented.

The feature set of creating and implementing a critical chain schedule therefore follows the path outlined below.

- Construct the network of activities using the estimated actual times, not elapsed, and with all proper logical connections made with specific reference to resources and the links between them.
- Ensure that the project team have the opportunity to consider the activities in terms of where the inputs are coming from, who is providing the inputs, the precise nature of the tasks, the expected outputs of the task, where the outputs have to go, the equipment required, and so on.
- Submit the network to critical chain software to identify the critical chain itself and to check for problems, firstly those of logical

connection and then of resource conflict. Allow the software to provide the initial schedule.

- Having determined that the critical chain identified meets the needs of the client, insert the appropriate buffers: project buffer, feeding buffers and resource buffers.
- Once more ensure that the overall project plan still complies with the requirement of the client and the organisation.
- Communicate the plan to all involved parties for review and comment to ensure that they can sign up to the plan and their commitment to it.
- Ensure that the changes to the measurement systems have been agreed and implemented especially with respect to resource effectiveness versus efficiency.
- Ensure that all parties to the project are able to work without interruption while performing their tasks and that they complete the task as soon as possible and then pass it on, i.e. report completion as soon as it is completed.
- Ensure that all suppliers work to the same rules and procedures, and delivery measures.
- Ensure that buffer management principles are properly understood with agreement as to roles and responsibilities should problems arise.

This set of features forms the basis for critical chain implementation. Once the project has been launched the focus moves from estimation to execution. The reporting of tasks now forms an integral part of the whole process. It is important that remaining task durations are supplied so that the project plan can be upgraded and buffer penetration determined. Now the estimations are of no value, it is simply sufficient to focus on the sequence, and move the tasks through as fast as possible keeping to quality on all other contractual requirements.

It is this process that the companies from which the case studies are drawn followed. In all cases they had made the decision to implement the critical chain approach on a range of projects. The case studies are drawn from companies in both civil engineering and new product innovation. They are companies of all sizes and from a number of geographical locations throughout Europe. The next chapter introduces the case studies in more detail and then we can assess the approach both in terms of project management, and also in terms of an enterprise focus. This will also stimulate the debate about how such an approach can be obtained.

CHAPTER FOUR

Case studies

Introduction

This chapter introduces the main case studies which form the basis for this book. They were drawn over a four-year period and come from the construction industry and the new product innovation industry. Although only two case studies are described in detail, others are used throughout the book for additional illumination. The first is drawn from the product innovation environment and the second from one aspect of civil engineering.

The current situation within project environments – case studies

This first case study is taken from a major player in the project-based industry with many projects on-going and with people working on more than one project at any time, in other words a typical multi-project environment. The people are part of a large multi-national organisation with an overall turnover in excess of $25 billion. The market in which they operate is subject to fierce competition and the penalties for being late into the market with a product can be severe. Throughout the organisation there is considerable pressure to maintain timely delivery of all projects, which in turn helps to maintain the viability of the organisation and its place in the market and thus to remain attractive to investors. The analysis was started after a period of about twelve months where the organisation had been active in the implementation of the critical chain approach. It had started in Europe with a small pilot implementation, which had gone successfully, although limited by the scope of the project itself. This success had led to the introduction of the critical chain approach in four main centres in Europe and across a product range. The numbers of people involved was some three hundred engineers plus associated support staff, planners and managers. The nature of the organisation, however, seemed to be preventing the real gains expected from the approach. Although individual projects were performing well, the global performance was below expectation. This led to a small group of senior managers from the main sites coming together to analyse just why the performance was below par, and to develop a solution that would substantially increase revenue, and improve the performance of the individual sites. The process started with them working to develop a set of UDEs to act as the raw material for the analysis. As these people held high positions they had a clear understanding of the problems, not just at local level but also at the senior level up to the board. The UDEs they collected were subjected to analysis prior to the programme as well as some scrutiny at the start of it.

The UDE list they developed included, in no particular order, the following set. There were more than is included here. This is just a sample reflecting the types and scope of UDEs considered.

- There are constant feature changes.
- Release dates are delayed.
- Costs escalate higher than budget.
- There is a lack of qualified support people.
- Internal delivery dates are not met.
- The project data is often incomplete.
- Some key project decisions are delayed.
- The organisation cannot absorb new contract features into the existing work programme.
- Work programme priorities are not communicated to project managers.
- Resource productivity is not meeting targets.

Although the UDEs were a selection of the total they were also the most common set produced by the internal team. The UDEs were subjected to scrutiny to ensure that they not only reflected the current environment but that, where possible, the financial implications were clearly understood. In this particular set of UDEs the total financial element was some $7 200 000 per year. This is the amount of money either being lost, or spent in order to cope with the continued existence of the UDEs. It was the recognition that this amount of money was involved that assisted with gaining the focus of the people involved in the analysis. This reflects the fact that often the true nature and impact of UDEs is not fully recognised until they have been both quantified and the links of causality determined. If ever further evidence was needed for the fallacy of only dealing with UDEs one at a time then this must surely be it. Equally, if as a result of the analysis it is possible to determine a cause for most of the UDEs within the current environment, then the gain is substantially more than that from any one UDE. In other words, through substantially enhanced focus, a major leverage on bottom-line performance can be achieved.

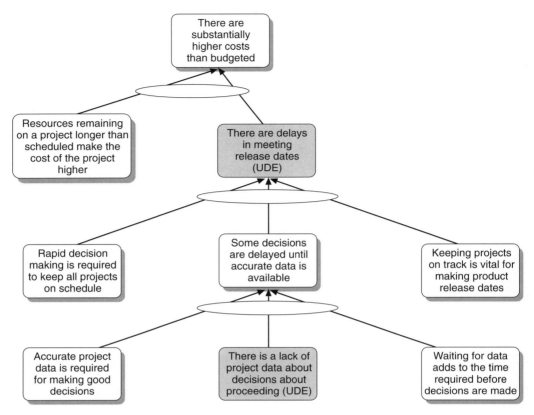

Fig. 4.1. Initial analysis of one UDE

One further analysis was carried out on a number of the UDEs, which simply asks the question just how bad can it get? Here the idea is to develop a simple logical analysis which shows what happens if the UDE is not addressed. This technique is called the negative branch reservation (NBR) within the TOC/TP set of tools. The idea is simple, just take one starting point, in this case an UDE, and ask yourself where this might lead. In combination with what other aspects of the current situation will the UDE lead to something worse? There is even the likelihood that there will be a connection to at least one more UDE.

This simple analysis started with one of the UDEs (Fig. 4.1) from the list and developed an argument based on the premise of leaving the UDE untouched, what will happen? It is easy to see that in a very short time a second UDE is connected to the first, and the inevitable outcome of higher costs against budget is reached. The person who constructed the logic then presented it to the others in the group for their comments. The result was confirmation that this was an excellent analysis of their

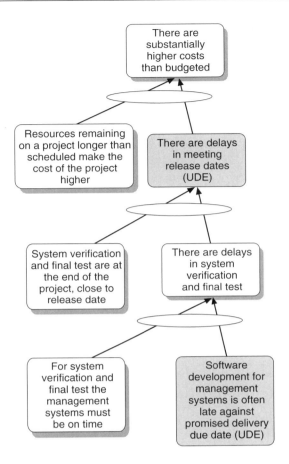

Fig. 4.2. Initial analysis of a second UDE

current environment and the recognition that what appeared to be a problem related to simply project specification at the outset was having such an impact on the project later in the day. This was an example of something, which was rather obvious, but the real impact had not been fully recognised before. They then carried out the same analysis on another UDE (Fig. 4.2).

Once more the end result illuminated what their intuition was telling them. The delays recorded here were not huge by any measure, they were just a little late, but the cumulative impact was often out of all proportion to the degree of lateness. This was really new to the team; the real impact had not been fully understood until this point. Remember, these are people who have a huge number of UDEs operating at any one time, they have to deal with them all at the same time, and that means they rarely have the opportunity or the time to properly analyse them. The results are not overly surprising, indeed they were no surprise at all to the team. The clarity of the causality was however. They began to see patterns in the connections between UDEs, including some UDEs they had initially felt were in no way connected.

Following the process of the TOC enterprise approach, once the UDEs had been properly validated and quantified, the next step was to construct the clouds that each UDE represents with the UDE cloud format. It is not usually necessary to build clouds from every UDE, typically three or four are all that is required in order to gain the necessary insight into what is currently going on.

This cloud (Fig. 4.3) represented a major problem within the company. They were continually under pressure to deliver, while at the same time trying to meet the necessary quality criteria that the market imposed. Often they found that the product failed to meet these criteria and that

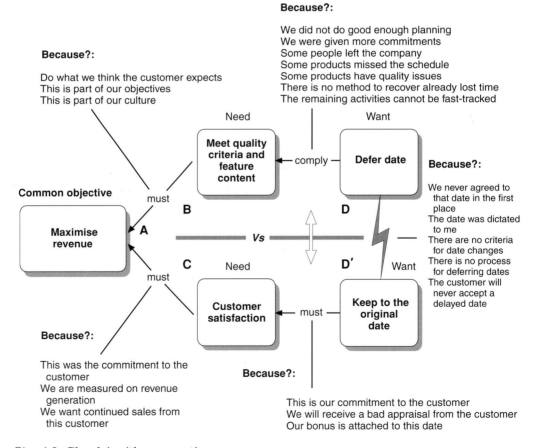

Fig. 4.3. Cloud 1 with assumptions

further work was required. This in turn required more time, hence the need for delay. However, both marketing and the other project managers, who wanted people released from this project, were keen to see no delay whatsoever. Project meetings often resembled battle zones as project managers and resource managers argued for hours over which projects could be delayed, which could not, and who could actually do the work. The lack of visibility across all projects was all too obvious. This one UDE alone accounted for a substantial amount of managerial time and stress. Of course the UDE was not alone, it had brothers and sisters to keep it company.

The UDE in Fig. 4.4 reflected the problem of the first UDE further down the line. The intention was to deliver the whole product by the completion

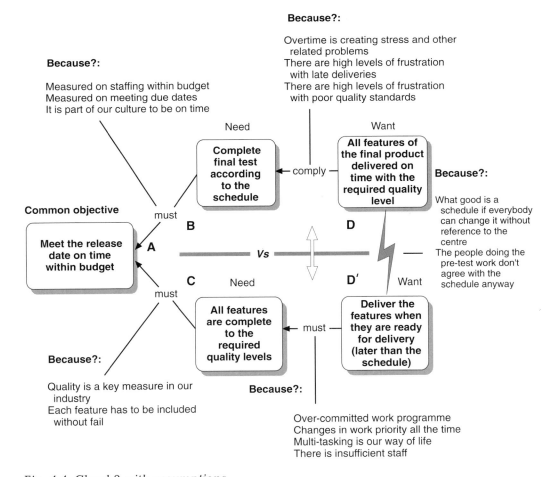

Fig. 4.4. Cloud 2 with assumptions

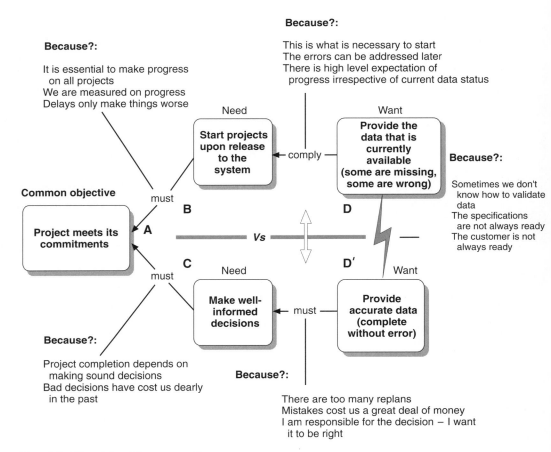

Because?:

This is what is necessary to start
The errors can be addressed later
There is high level expectation of
  progress irrespective of current data status

Because?:

It is essential to make progress
  on all projects
We are measured on progress
Delays only make things worse

Need

Start projects upon release to the system

Want

Provide the data that is currently available (some are missing, some are wrong)

comply

Because?:

Sometimes we don't
  know how to validate
  data
The specifications
  are not always ready
The customer is not
  always ready

Common objective

must

B

A

Project meets its commitments

Vs

D

D'

must

C

Need

Make well-informed decisions

must

Provide accurate data (complete without error)

Want

Because?:

Project completion depends on
  making sound decisions
Bad decisions have cost us dearly
  in the past

Because?:

There are too many replans
Mistakes cost us a great deal of money
I am responsible for the decision – I want
  it to be right

Fig. 4.5. Cloud 3 with assumptions

date agreed with the customer. However, it was clear that there were many times when the full feature set was simply not ready for release. The project team usually argued for releasing what was available as they felt that this kept the customer happy among other things, but marketing, and often the customer as well, wanted to have the full feature set, now!

Cloud 3 (Fig. 4.5) focused on a further issue that was concentrating minds, ensuring that the data required to make decisions was both available and accurate. Many of the project meetings were held with often scant information about progress, availability of people and so on. Decisions still had to made, however, so they were, even if the people making them knew that the decision would come back again in a short period of time for both review, and probably change. This gave rise to some very interesting communications taking place between functions

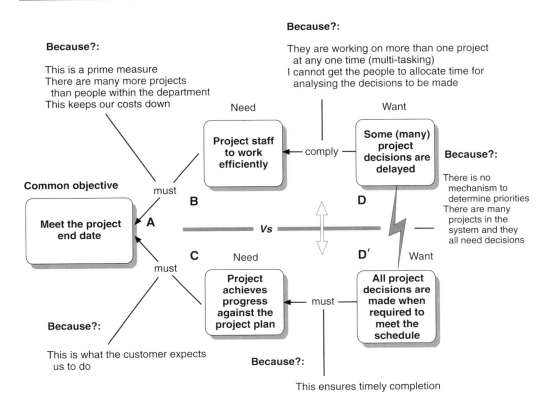

Because?:

This is a prime measure
There are many more projects
than people within the department
This keeps our costs down

Because?:

They are working on more than one project
at any one time (multi-tasking)
I cannot get the people to allocate time for
analysing the decisions to be made

Need

Want

**Project staff
to work
efficiently**

← comply —

**Some (many)
project
decisions are
delayed**

Because?:

There is no
mechanism to
determine priorities
There are many
projects in the
system and they
all need decisions

Common objective

must

B

D

**Meet the project
end date**

A

Vs

**Project
achieves
progress
against the
project plan**

— must —

**All project
decisions are
made when
required to
meet the
schedule**

must

C

Need

D'

Want

Because?:

This is what the customer expects
us to do

Because?:

This ensures timely completion

Fig. 4.6. Cloud 4 with assumptions

and levels within the organisation as to what was really happening! It has to be said that the stress levels were often high during these meetings and the blame culture of the organisation was frequently evident.

Cloud 4 (Fig. 4.6) is an example of the earlier delay cloud although with a slightly different focus. This cloud was driven by the measures in use to determine resource activity. To many resource managers it did not matter what projects people were working on, as long as they were working on at least one all of the time. If they were bouncing between projects, continually filling their time then that was fine. The measurement system did not distinguish between project priorities, only if people were idle would the system raise a question. Filling the time sheet was a regular excursion into an unknown world, one of almost fictional qualities. Thus it was common for people to work on the wrong project, not be available when they should have been, and above all insert substantial delays into the decision-making process.

It was this cloud that highlighted the fallacy of using efficiency measures as a reasonable measure of individual performance. This was

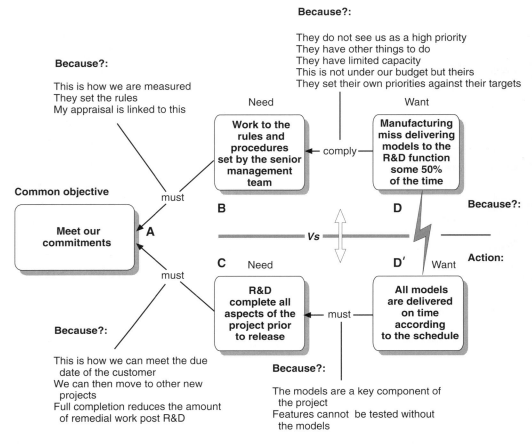

Because?:

They do not see us as a high priority
They have other things to do
They have limited capacity
This is not under our budget but theirs
They set their own priorities against their targets

Because?:

This is how we are measured
They set the rules
My appraisal is linked to this

Need

Want

Common objective

Meet our commitments   A

must

B

Work to the rules and procedures set by the senior management team

← comply —

Manufacturing miss delivering models to the R&D function some 50% of the time

D

Because?:

Vs

C   Need

R&D complete all aspects of the project prior to release

← must —

D′   Want

All models are delivered on time according to the schedule

Action:

must

Because?:

This is how we can meet the due date of the customer
We can then move to other new projects
Full completion reduces the amount of remedial work post R&D

Because?:

The models are a key component of the project
Features cannot be tested without the models

Fig. 4.7. Cloud 5 with assumptions

very hard for the team to come to terms with. Efficiencies were part of their culture and now they could see how chasing efficiencies was damaging overall performance.

The final cloud (Fig. 4.7) focused on a different aspect to those clouds already discussed. For every product it was necessary for the manufacturing function within the company to produce models for checking dimensions, fit, etc. for the final product after the hardware had been designed and produced. Without the models it was difficult to determine final fit, and therefore the project could not be signed off. Often the R&D people would be called back into manufacturing to resolve a problem with the product after it had left the R&D function itself. Apart from the obvious impact this had on production schedules and the release of the product into the market, it also meant that key project staff had to carry

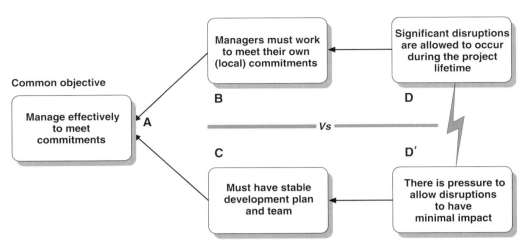

Fig. 4.8. The composite cloud

out what was effectively remedial activity and leave their current work until they had resolved the issue.

Once the five clouds had been properly constructed and the assumptions determined, the next step was to construct a composite cloud that drew on all five but which retained the power and insight of each. This was achieved by laying each cloud side by side, and looking for the patterns in each one that could lead to a unified, composite cloud. It did not take the team long to construct the cloud shown in Fig. 4.8.

Once the cloud had been constructed it was then checked by the group. Did it capture the real sense of the conflict? Did the cloud fall into line with the intuition of the team? Did each member of the team have first hand experience of the conflict, and the problems in trying to deal with it? How real was the analysis? It did not take long for full agreement that the cloud did capture the conflicts, did fall into line with their intuition, and they all had stories which demonstrated the on-going struggle to resolve the conflict, that this was all too real. Once this level of agreement had been achieved the next step was to surface the assumptions that lay behind each of the arrows. Each person did this on their own, sometimes returning to the original set of clouds they each had to see if any assumptions from there also fitted here. Remember they each represented a different location, in a different country, and this ability to verify that the same problems were common to all four locations gave enormous confidence that they probably operated in the other locations as well. Ensuring that the other locations had the opportunity to validate the logic analyses was central to their gaining buy-in in the other sites.

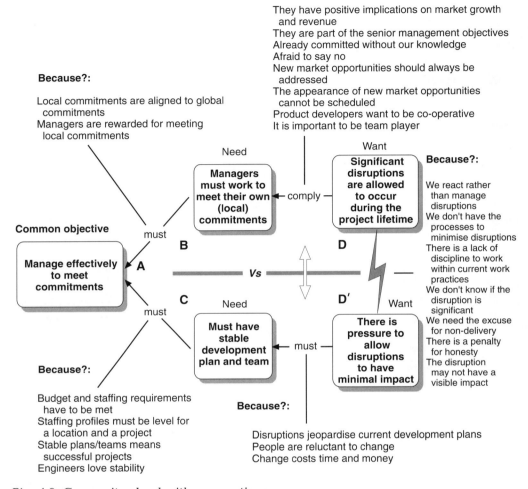

Because?:

They have positive implications on market growth
and revenue
They are part of the senior management objectives
Already committed without our knowledge
Afraid to say no
New market opportunities should always be
addressed
The appearance of new market opportunities
cannot be scheduled
Product developers want to be co-operative
It is important to be team player

Because?:

Local commitments are aligned to global
commitments
Managers are rewarded for meeting
local commitments

Because?:

We react rather
than manage
disruptions
We don't have the
processes to
minimise disruptions
There is a lack of
discipline to work
within current work
practices
We don't know if the
disruption is
significant
We need the excuse
for non-delivery
There is a penalty
for honesty
The disruption
may not have a
visible impact

Because?:

Budget and staffing requirements
have to be met
Staffing profiles must be level for
a location and a project
Stable plans/teams means
successful projects
Engineers love stability

Because?:

Disruptions jeopardise current development plans
People are reluctant to change
Change costs time and money

Fig. 4.9. Composite cloud with assumptions

The surfacing of the assumptions (Fig. 4.9) was a revealing exercise for the group. They had really to come to terms with some of them, and in some cases realise that they were part of the problem. They also started to realise the systemic nature of the issues, a realisation that is very much in line with the observation of Deming that most problems are systems driven not people driven. Although it has to be said that people were certainly responsible for maintaining some of the problem issues in the way they were managing the project environment. They then checked, and rechecked the logic of the cloud and the assumptions in preparation for the next step.

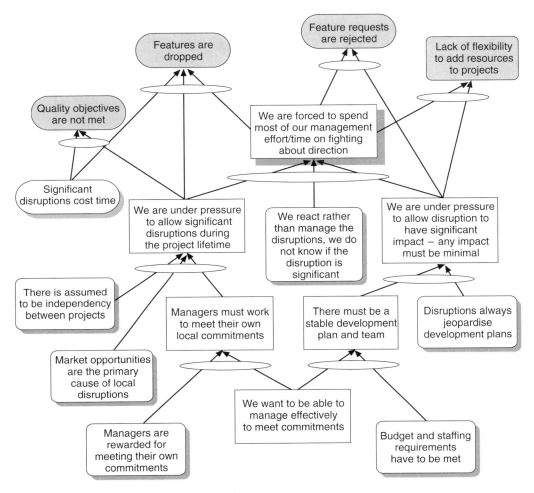

Fig. 4.10. The communication analysis

The composite cloud once validated allows for the construction of the communication analysis (CA), which appears in Fig. 4.10.

This analysis forms the basis for communicating the work they had done to others within the company, in both their own sites, and those other sites not included in this team. The team spent a great deal of time checking the logic of this analysis. They also wanted to make sure that they could connect to other UDEs not yet included in the overall analysis. By this time they had become convinced that they were close to understanding the real causality within their organisation, and how it was driving the many UDEs they experienced day after day.

Once this whole analysis had been validated in the class, the next step was to check the analysis with others in the field. This aspect took a few

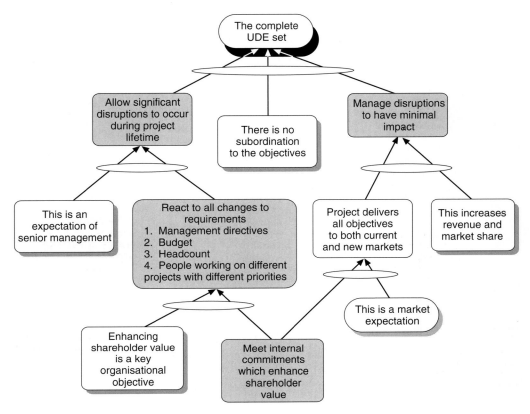

Fig. 4.11. Second communication analysis

days and gave full confirmation that each step of the analysis was sound and that the core driver for the UDE set was as indicated. At this point it was agreed that consensus on the problem had been achieved, and thus the first step of the TOC change model had been achieved.

Despite this the group wanted to gain further confirmation of their analysis. They developed a second composite cloud of which the communication analysis is shown in Fig. 4.11.

This second analysis was used for communication to the senior management team in order to enable them to see the impact of the core drives being out of alignment. The team knew that they had to enlist the support of the senior management team if they were to have any impact on the organisation. This meant that they used different parts of their whole analysis depending on who they were going to present it to. They recognised that the argument that might capture the shop-floor would not necessarily do the same for the top team, and vice versa. Gaining the buy-in of the others within the organisation was recognised

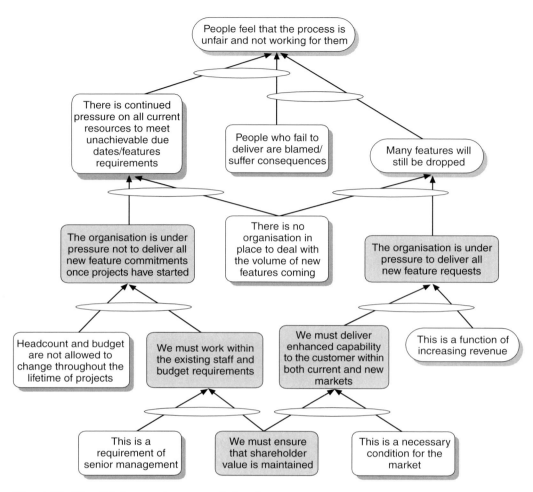

Fig. 4.12. The CA for engineers and developers

as fundamental to a successful conclusion – in other words ensure that steps two and three were achieved, consensus on the direction of the solution, and consensus on the benefits of the solution. To that end one further CA was developed (Fig. 4.12).

This CA was developed for communicating to the people who had already bought in to the critical chain approach as a conceptual solution, and had tried to implement it in their own areas. They had, however, found that that was insufficient to really deliver the kind of results they had expected. Here they saw that part of the problem lay in the systems currently in place to try and deal with the many problems they had experienced prior to the critical chain implementation. The systems, although designed to function in a pre-critical chain environment, were still in place, they had not been changed or challenged in any way, and yet the reasons for having them had been addressed by critical chain.

Developing the solution was the next step. This involved the construction of first the core future reality tree (CFRT) and then the full implementation plan.

The feature set of the solution

Once the base of the solution had been constructed and validated through the use of the Core FRT, the next step was to determine the feature set of the full solution. This involved returning to the composite cloud and examining the assumptions surfaced there. It also involved the integration of comments and suggestions from the rest of the team. This led to the creation of the following set.

- The implications of market opportunities on both market growth and revenue are properly evaluated and validated.
- There is a reward system for all team members based on risk taking and achieving aligned goals.
- All goals throughout the organisation are aligned with those of senior management and the organisation as a whole.
- Unilateral decisions are eliminated.
- The person making a commitment is responsible for that commitment.
- We have business data that shows profitability versus risk to other parts of the business unit and the other projects.
- There is sufficient capacity to meet the expected market demand.
- There is a process for handling unscheduled inputs (disruptions).
- Product developers only provide information to the decision-making process for taking on new products.
- Risk analysis for all disruptions is carried out.
- Tools for determining current status are in place.
- Rewards and penalties are in place to encourage following the current written practices.
- Senior management support the discipline to work with the current practices.
- There is a culture to accept responsibilities for our own mistakes and not *shoot* people for making a mistake.
- There is a culture of not punishing people for telling it like it is.

This set of features was developed by the team in readiness for taking full advantage of the critical chain implementation that was on-going.

Once the implementation team had been set up and the process rolled forward it soon became noticeable that progress was still not forthcoming. A series of meetings were held where further problems were highlighted.

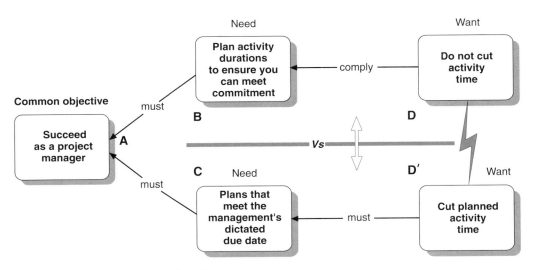

Fig. 4.13. Management cutting times

Although solutions had been developed, although it was agreed at the highest level, although the engineers and developers had all bought in to the process and the plan, they still did not see the benefits. From this series of meetings three clouds were developed. They all shared the same basic issue, management involvement. The three clouds are shown in Figs 4.13–4.15.

What the cloud in Fig. 4.13 led to was that times were cut without adding the buffers. Although the management team fully understood

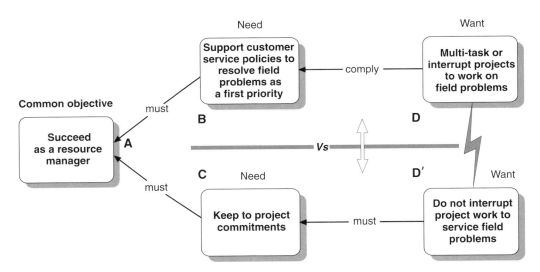

Fig. 4.14. Allowing multi-tasking to continue

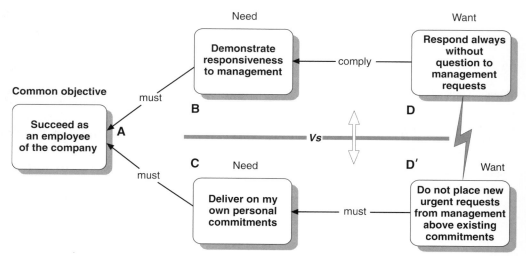

Fig. 4.15. Dealing with urgent requests

the reasons for the times as they were being prepared, they still tried to cut into them, and the buffer time, which was the overall protection. The pressure on the company from the market was such that they felt it necessary to do this. If this was not bad enough the old problem of multi-tasking still continued to create problems for the whole team.

The cloud in Fig. 4.14 allowed people to continue to permit multi-tasking even though they knew the impact that would result, and did. This was as much a function of people still not being comfortable with saying 'no' when the interruption came. This lack of focus, this lack of sub-ordination was a source of some frustration among the implementation team. This led to the final cloud of the three (Fig. 4.15).

People felt part of a team and that meant, to them, that they should always respond without question. The fact that this had a major impact was known, but ignored. Constructing the cloud allowed people to really see the impact, and the real effect it was having on the overall project team and the project itself. There was no mechanism to check the validity of any urgent request, many of which turned out to be less than important and could well have waited.

The three clouds were then submitted to composite cloud construction to see the overall impact of each of them. This resulted in Fig. 4.16.

Once this cloud was constructed and checked, it became clear that the core of the problem lay in two areas. The first was that we now had a conflict of subordination to deal with, and the second was that while we had been successful in implementing a new approach for projects and project management we had failed to challenge and implement changes

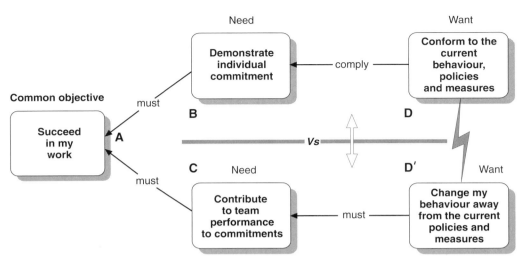

Fig. 4.16. *The power of existing rules and procedures*

to the measurement process and the policies, rules and procedures that surrounded it.

When this analysis had been completed a solution was developed and integrated with the original one developed previously. The current situation is that the company continues to develop the use and application of both the critical chain and TOC management methodology. Although there is still a great deal of work to do, the analysis to date has shown the typical sets of problems faced in this industry, and the level of thinking that has to be done for any real progress to be made in terms of overall performance on the bottom line.

The second case study

This case study was developed from work completed over a nine-month period prior to the Hatfield disaster with a major construction company working within the rail industry. The study started with a period of data capture where we were asked to visit a number of rail depots and determine the type and scale of the problems they were facing. The questions asked centred on collecting the current set of difficulties being experienced by the people on the sites around the region. This data was captured through a series of informal interviews with both front-line managers and engineers. These included both permanent way and signalling teams/managers. With the senior management team, time was spent preparing the data for a full analysis to determine the core drivers within the rail environment, and what the real issues were. Over 66 UDEs were validated as a result of this activity.

The company used a project-based approach for the successful running of all the maintenance, repair and upgrade work it was responsible for. This whole data capture process took about eight weeks to complete and involved five people. The intention was to capture the widest range of UDEs possible from this environment and use them as the basis for an enterprise analysis using the TOC/TP. Seven depots were involved in this activity, and between 15 and 18 people took part in the data capture from the company.

Once the UDEs had been collected they were then subjected to scrutiny and a short list of some 25 UDEs were used as the basis for the initial analysis. The process of reduction took about five hours. There was a great deal of duplication in the UDEs, but there were also UDEs that reflected individual grievances which were more of an interpersonal issue than organisational and these too were removed from the final list. There were a small group of people attending the initial analysis sessions and they were asked to choose a number of UDEs from the list that feed into their area of control and influence. Remember it is important that people address those issues under their control where they have the responsibility to do something about it. Soon there was a list of UDEs that each person felt was representative of the environment and which comprised a substantial number of the problems that were costing the company a great deal of time and money. Part of the list is shown below.

- There is only limited flexibility between the teams.
- Maintenance work is not effectively planned.
- There is insufficient resource to cover emergency and maintenance work.
- Often previous maintenance repairs have to be redone.
- On-track plant is not always available for possession work.
- Jobs are not always carried out effectively.
- Site teams do not have targets to work to.
- We are often unable to complete jobs within possession time.
- The performance of the team is inhibited by inadequate support systems.
- Confirmation of plant availability is often late.
- People do not know what their contribution is to the business.
- No measures exist for track quality.
- Some people are not skilled enough to produce work to specification.
- The data for making proper prioritisation decisions is not always available.

There were more, but for the purposes of this book this is a representative list. Each UDE was then examined to determine the impact it had on its own on the performance of the organisation. This was set into the context

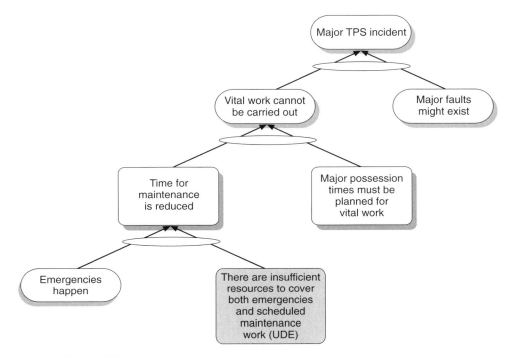

Fig. 4.17. First UDE analysis

of the key performance indicators (KPI) required by Railtrack, and also of the parent company. These KPIs are as follows – safety, track quality and train performance. They were also determined in both financial and non-financial terms. Having spent some time validating the UDEs and their impact on the overall performance of the company, the next step was to consider what happens if the UDEs are left alone, in other words if nothing is done to alleviate them either individually or as a whole. The tool for this is the NBR and one example is shown in Fig. 4.17. This analysis was already clearly understood by the team and they could cite a number of examples where it had actually happened.

Once the impact of doing nothing had been accomplished, the next step was to analyse the problems the continued existence of the UDEs had on the interpersonal relationships within the organisation. UDEs will create conflicts between people and between departments. In time these con-flicts can themselves have a major impact on the ability of the organisation to function. Recognising that it is the continued existence of the UDE that creates the conflict is further evidence of Deming's observation about systems and problems discussed earlier. Each of the team was asked to pick two or three UDEs and describe, using the conflict cloud technique, the kinds of conflicts the continued existence was having on them and the

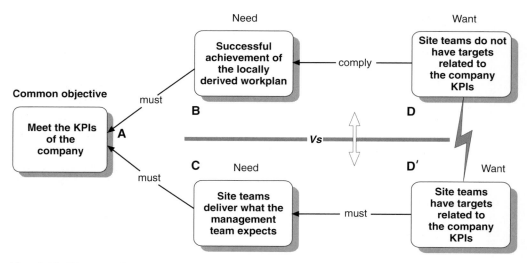

Fig. 4.18. First conflict cloud

people they worked with. A couple of examples are shown in Figs 4.18 and 4.19.

The cloud in Fig. 4.18 was a function of the need to align the work and the decisions made to the overall objectives of the region. Site teams had not, in the past, had clear direction from the centre. They had become used to developing their own approach to the railway and operating accordingly. Many had experienced numerous changes of corporate identity in that time and felt that they could outlive the current operators.

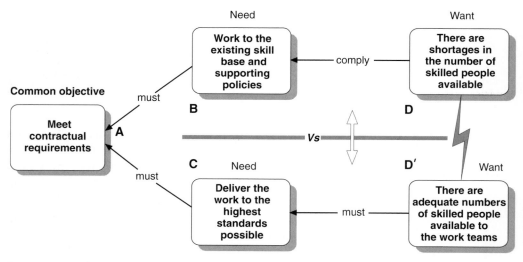

Fig. 4.19. Second conflict analysis

This meant that often the work content of possessions was not in line with the overall needs of the network. The project nature of the work was lost in the myriad work elements being carried out which in turn made clear management very difficult. Meetings held between the site people and the management team often centred on the mismatch between the targets and the need to resolve the differences relatively quickly. Often meetings would degenerate into minute detail about which part of a target was important, and which was not. Local people felt that their independence and flexible approach was being attacked and required defence. The management team felt that they had to gain control over the various teams in order to provide a coherent and focused response to the whole maintenance question. Although these two elements are not in conflict, the way in which either side wanted to achieve them was most certainly in conflict.

The cloud in Fig. 4.19 was also a function of the current environment and once more related to the history of the industry. Many highly skilled staff had decided to vote with their feet and leave the industry. As a result many depots were struggling with reduced numbers of highly skilled people. As the pressure to meet the new contractual demands also came into play, there were restrictions on recruitment, which only served to exacerbate the problem.

The next step was to start to analyse the UDEs themselves using the UDE cloud approach. The process being followed started with each person constructing three UDEs clouds, and then having each one scrutinised (Fig. 4.20).

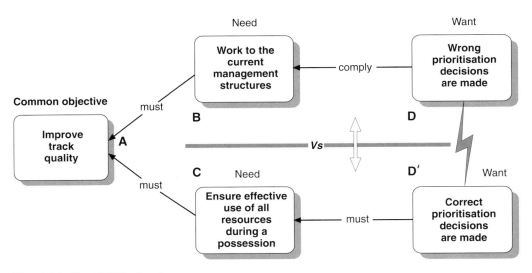

Fig. 4.20. First UDE cloud

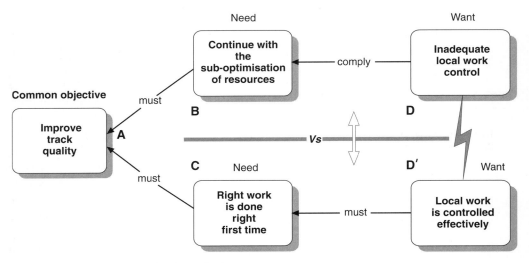

Fig. 4.21. Second UDE cloud analysis

Given the issue raised above about the way in which local teams could do their own work routines, it is not surprising to find that wrong decisions were being made. Often the teams and their supervisors lacked the skill and knowledge to make the proper decisions. Often the guidance coming out of the centre failed to give the right level of direction. The result was never in doubt, however. Many times work was carried out which could easily have waited, tying up scarce resources and allowing other, more important problems to remain unresolved.

The cloud in Fig. 4.21 only serves to emphasise the issues already raised. Once more the problem revolves around the control at local level and the way such resources are allocated to work. Given the scarcity of resources, not just people, it rapidly became a key issue to address.

The final cloud (Fig. 4.22) concerned an issue of some importance, the way in which faults were reported. The people responsible for reporting faults felt it necessary to protect themselves by raising the priority of every fault to the same level – urgent. This placed a substantial load on the system. Again, part of the problem lay in the person doing the analysis. Being part of the local team they were concerned to ensure that their team did not come under the fierce scrutiny that accompanied making the wrong decision.

Once the clouds were completed, the next step was for each person to construct a composite cloud from the three UDEs they already had. Once each person has a composite UDE cloud they were then combined into one generic cloud, which brought together all the clouds of the managers from the company (Fig. 4.23).

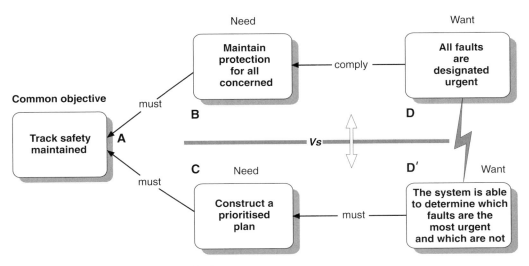

Fig. 4.22. Third UDE cloud analysis

This cloud reflected the real problems in the industry in terms of its recent past, the lack of investment over the previous few years, the uncertainty over the future for many of the people working on the ground and so on. Although the team was only looking at one small part of the whole network, they quickly saw the same problem in other parts. Many of the people working at ground level simply had no visibility as to how they fitted into an overall structure, an overall strategy.

Once a scrutinised composite cloud had been constructed the next step was to build the CA drawing on the assumptions that had been surfaced

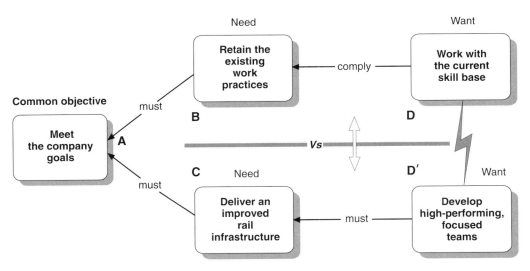

Fig. 4.23. Composite cloud

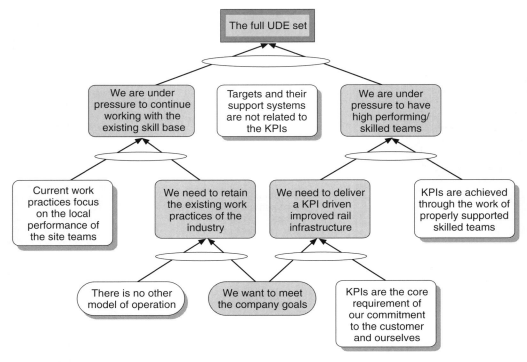

Fig. 4.24. The communication analysis (CA)

throughout the whole process to date. If the analysis held true then this would satisfy the first step, that of consensus on the problems facing the whole of the organisation. The CA is shown in Fig. 4.24.

Once agreement had been reached on the CA, and time spent back in the rest of the company checking the validity of the analysis to date, the next step was started, that of gaining consensus on the direction of the solution. This began with identifying the DEs that corresponded to each UDE and checking them for both clarity and a suitable measurement. This last aspect is important if there is to be confidence about the achievement of the DE itself.

At this point a comparison was made between the traditional approach that was being used for the management of projects within the company and the critical chain approach. This was done to highlight the difference and provide people with a quick and simple method of comparison. I am grateful to Phil Bayliss for the idea and the construction of Table 4.1.

This simple analysis was used with great effect to make a sound comparison between the two approaches of critical path and critical chain. Once the communication exercise was completed in terms of the initial solution the full feature set was developed.

Table 4.1. Comparison chart between critical path and critical chain

Traditional style of projects	Critical chain style of projects
Individual behaviour	*Individual behaviour*
Instinctively we avoid things damaging to us	Open and honest culture
Working for oneself	Working together
Blame culture	No blame
Network planning	*Network planning*
Estimate of task durations at low risk to the person – 80% probability	Estimation of durations – 50% probability
Critical path – longest line of task dependent events	Network validation, task and resource definition
	Identify the critical chain – resource dependent as well as task dependency.
	Buffers included at strategic positions
Task behaviour (execution)	*Task behaviour (execution)*
Milestone/date driven	Only carry out planned tasks when required
Multi-tasking is the norm	Execute the task as quickly as possible
Urgent task demand	Focus only on that task until completed – prevent multi-tasking.
No reporting of early finishes	Report remaining duration to completion
	Report early finishes
	Buffer management for all buffers
Effect on the project	*Effect on the project*
Late, missed original due date	On-time or early completion
Overspend on budget	Within the budget
Specification cut back to meet time targets	Controlled focus as to when and by how much to spend
Firefighting continuously	Meet the specification
Low morale	Reduction in overall hours per project
Continual changing of priorities	Higher safety standards met
Lower margin than forecast	Reduced firefighting
	Higher morale
	Margin substantially increased
Effects on the organisation	*Effects on the organisation*
Rapidly becoming uncompetitive	More competitive
Turnover targets not reached	Higher market share
Profit targets not reached	Achieve turnover target
Share value reduction	Achieve profit target
Effects on the client	*Effects on the client*
Unhappy client	Happier client
Takes future business to the competition	Repeat business
	Able to choose clients

The feature set of the solution – second case study

- Targets exist that relate to key performance indicators (KPIs) with systems that support meeting those KPIs.
- The skill base is developed in line with work targets.
- The teams focus on local targets derived from the KPIs.
- The KPIs and progress towards them is constantly communicated to the whole organisation.
- The local management teams are assessed and developed to deliver the local KPI driven targets.
- A system for detailing work performed, with costing and resultant condition of track included, is implemented.
- A framework of financial control is implemented across the whole region.
- Prioritisation of work is accomplished based on an enterprise wide information system.

This feature set formed the basis of the solution, the implementation plan constructed and implemented. The work began and although there were a number of teething problems, the implementation delivered improved performance for the region.

Initial conclusions from these case studies

The case studies included here represent a good cross-section of the projects undertaken during the time of the research. What became fascinating was that when a group of people from civil engineering were presented with the UDEs from R&D but without knowing they were from that environment, they felt that they were a good represen-tation of civil engineering UDEs. The opportunity to do the same with civil engineering UDEs with a group of people from R&D was also undertaken with the same end result. The fact that the UDEs were in fact almost all related to management issues rather than industry specific issues undoubtedly helped, and served to demonstrate that the primary issues in project management are generic in nature.

The lack of focus was common throughout all the case studies, not just the two highlighted here. There was often a high degree of focus at a very micro level, but the overall focus related to the performance of the whole was by and large missing. This led to many decisions being taken at too low a level within the organisation and without sufficient understanding of the wider implica-tions of any one decision. This in turn led to considerable duplication of work, missed opportunities and high levels of frustration at all levels.

On occasions, the clouds were used to show to other companies with the view of gaining their acceptance of this approach to their own problems.

The new problem that resulted was that in so doing they thought we had already examined their company and wanted to know when we had carried out the research. This was, of course, very welcome feedback.

Throughout all the case studies it was the management issues that dominated. The lack of visibility of the revenue chain and the lack of measures to determine progress towards the goal of the organisation were common problems. The inability to successfully manage change was a constant theme, likewise the inability to subordinate to the decisions that had already been taken, and agreed with all parties.

Alhough some groups were highly successful with their individual implementations, the whole organisation was still struggling to make the best out of the process being adopted. This led to the recognition that even when the solution is fully agreed, understood and implemented in a positive manner, it can still fall short if the organisational context is not addressed at the same time.

The rest of this book will examine some of these issues in the drive to create a real enterprise focus linked to the ability to implement change successfully within the organisation. The case studies have served the purpose of demonstrating that the key issues in project management lie neither in the content of the project plan, nor in the structure of that plan, but in the management and associated measurement structures that apply. The critical chain approach to project plan development and management has shown itself to be more than capable of delivering projects on time, to budget and without sacrificing specifications. However, to be really successful it demands a change to the fundamental management structure operating within the organisation. This is where our focus now turns.

CHAPTER FIVE

Enterprise project management through the application of the TOC

Introduction

This chapter introduces the concept of the enterprise focus for project management which lies at the heart of this book and which I feel is the only way forward for project-based companies. Those who fail to heed the warnings may not survive in the long term, or even the short, and the competition will take over. Those companies who feel secure today may not feel so secure tomorrow. Although the pressure to change may not be visible right now, that pressure will become more and more insistent over the next few years. The nature of contractual relationships has changed substantially and the market expectation has also changed. The future is clearly heading towards enterprise solutions in terms of resources and productivity. It is also heading in that direction as a strategic requirement as more and more companies recognise the reality of working the supply chain properly. The level of interest in supply partnerships, developing supplier relationships and so on demonstrates a desire to at least recognise the importance of the supply–client issues.

Ptak and Schragenheim (2000) when considering an enterprise resource planning (ERP) approach to organisations argued that

> *There should be a synergy between the various parts of the organisation, otherwise there is no value in staying together in one organisation. Behind this synergy lies the need to synchronise many different resources and activities in order to create value in the eyes of the customer. The synchronisation of the various parts generates dependencies within the organisation.*

They go on to argue that 'It is not feasible to manage so many dependencies and variables such that the products and services will achieve customer satisfaction while utilising many resources to their maximum potential level'. This argument is very much in line with the notion of the organisation as a chain, a series of dependent functions and activities. It also recognises that the ability of the chain to deliver is limited by the weakest link, the constraint, and that therefore running all functions to their maximum potential is simply wasting valuable resource time and money to no effect on the bottom line of the organisation itself.

Again Ptak and Schragenheim argue that

> *This emphasis on the global systems as a whole makes the TOC management accounting very different from marginal costing. What happens when the additional order has to be partially processed using overtime? In such a case when there is a direct link between the order and the overtime, the additional costs are to be calculated for the decision itself: ΔOE. The TOC decision rule is: $\Delta T - \Delta OE$ needs to be positive. This difference represents the additional profit, what is going directly to the bottom-line.*

Although using an example from production, the same can be said for projects. Where the buffer management demands a decision, the same rules apply as for production. In this way financial decision making is determined by the reports generated by the buffer management system which in turn is highlighting issues related to the key performance indicators of projects, due date completion, budget conformance and specification adherence.

Seeing the organisation as a whole is therefore a key component of an ERP system. Equally, from a TOC focus the same requirement applies. It is simply not possible to judge the impact of a decision, if the level of visibility is restricted to one area or possibly two at most. The level of visibility must be the whole revenue chain. On this point, Ptak and Schragenheim consider that

> *TOC speaks of an 'exploitation of constraints' and 'subordination to constraints' as necessary elements for properly managing the daily activities of the organisation. These two terms have a very strong impact on the role of the information system within the organisation. Exploitation means ensuring that the weakest links in the organisation are fully utilised and what is the most profitable for the organisation as a whole. Subordination means to design the processes throughout the organisation so that every process supports the exploitation scheme in the best way possible.*

This is a fundamental issue. The five steps of focusing of the TOC demand proper exploitation and subordination.

Within the two case studies of the previous chapter there was evidence of managers failing to properly exploit the project schedules at their disposal, failing to subordinate to the project constraint (the critical chain) and work in other areas. The continued acceptance of multi-tasking is the best example. Many managers failed to ensure that engineers worked only on the critical chain, often pulling them away to do other work without really thinking through the impact of that decision. Of course there will be times when such action is both necessary and correct, but the fact remains that in most cases no analysis took place, simply a knee-jerk reaction to an urgent demand without questions being asked. In many other organisations the same lack of focus and discipline were evident. In one case the ruling measurement was that of resource efficiency which, in that particular case, led to people being continually unable to start critical chain activities as they had been given other tasks to complete prior to the critical chain activity. The reason usually given was that the resource managers could not afford to have people waiting for the task to start, therefore they had to be given something to do. This was almost always done without reference to the overall performance of the projects, but it did make the resource managers look good when it came to resource activity measures. The system did not care which activities they did as long as they were continually allocated to activities, and often more than one at any one point in time. However, there are examples of companies who have been able to construct a wider focus, an enterprise focus.

Before embarking on the path towards an enterprise-focused organisation it might be better to assess the current situation first. What follows is not an exhaustive set of issues and questions to be considered, but this assessment certainly covers most of the key aspects in terms of organisational readiness.

First, what steps have been taken to shift from internally-focused local measurements to externally-focused global measurements? Do the measures in use determine progress towards the enhancement of the organisational revenue chain?

Second, what education has been carried out to shift the thinking of all the people from local to global, to enterprise wide?

Third, have the team dynamics of the people within the organisation been analysed and developed?

Fourth, have the financial people in the organisation been prepared to base all decision making on throughout accounting rather than cost-based measures? This has to include the purchasing department who can, by following the wrong set of measures, have a substantial and negative impact on overall performance.

Fifth, has appropriate software and hardware, designed to act as the delivery mechanism of information to all parts of the organisation, been properly specified and implemented? This system forms the basis for all information flows within the organisation and is therefore a necessary condition for the success of an enterprise focus.

Sixth, have all the appropriate systems been changed in order to show the interdependence of functions and performance, and the systemic nature of the organisation? This is a vital dimension of a successful enterprise approach. If the technology is implemented but the supporting systems remain as they were it is inevitable that the expected results will not follow.

Seventh, are all systems, all functions, all departments, all the people focused on the customer and the true nature of the supply chain?

Eighth, have all the rules and procedures been challenged and changed if necessary to ensure that the behaviour of all the people follows an enterprise focused pattern?

Ninth, how does your quality control system and supporting activities compare to world-class standards?

Tenth, has top management accepted and implemented the principle of creating a learning organisation, from top–down?

This is not a full list, there are many other items that could be added, but in my opinion this is a sound starting point. To implement this feature set is no easy task, no quick fix, but each of them will bring enhanced opportunity for all within the organisation. Let us examine an example of applying many of these features into a company with one more case study. This is taken from the high-tech industry, a company, and an industry, that has for some time recognised the need to innovate or risk oblivion. Other industries such as civil engineering should take note of this point.

An example of the enterprise focus using the TOC within a project-based industry

This example is of a high-tech company with the stated aim of being the 'leading high quality provider of technology-based products that enable people to get their information when, where, and how they want it – anywhere in the world'. This is a company involved in the design and manufacture of electronic devices such as disc drives, flash memory, other electronic components and associated software. It has over 60 000 employees in over 25 countries and considers itself to be the market leader in disc drives where they manufacture over 100 000 per day and have around 23% of the world market. They have some 30% of the tape drive market and their revenue is of the order $6·4 billion for the year 2000. They consider themselves to have world-class manufacturing and design capabilities.

Prior to the use of the TOC approach the environment in which products were undertaken was of some 12+ drive products being developed simultaneously, with no specific management tool being used. Key suppliers were not aligned with the programme requirements and there was a 2 × date system for communicating progress, one internal and one external. At the same time, which was around 1997, it was realised that their market position was being challenged. This led to a desire to move away from an 'execution-centric' focus to one of sustaining technical and process excellence. They recognised the need for change and put in place teams charged with examining the way they were doing things and to come up with innovative solutions.

With the introduction of core teams given the goal of increasing the decision-making bandwidth and the velocity by flattening out the decision-making structure, they started to run each product's development as a business unit. They added responsibility for finance and leadership authority to the core teams to execute. They established areas of co-location for teams in order to meet their needs and to provide weekly executive management reviews on progress and interventions. The teams comprised a small number of people with complementary skills who were committed to a common purpose, performance goals and the approach for which they held themselves mutually accountable.

Within the drive team the vision was to qualify the best drive for the high performance market ahead of the competition. Their mission was to work together as a proactive team and focus the resources necessary to develop and qualify a new high performance drive platform, which met the customer's and their own expectations. The market, as they saw it, had four main priorities: first, performance with a 25–30% increase in input/output/second using 15 000 rpm. The next was time to market with a production release in the fourth quarter of the financial year 2000. The third was cost where the target was to minimise cost delta with the mainstream 10 000 rpm drive and the final target was in acoustics. The most important of these priorities was that of time to market. In order to achieve these targets they opted to use the critical chain approach and the associated management methodology. The team started with the conceptual education programmes in order to fully understand what critical chain was about and to highlight any problems with implementation. They then began to work on network creations and validation with reviews before transferring the networks to the software they had chosen which was *Concerto*.

The initial network for the drive project consisted of some 180 tasks of which 36 were on the critical chain. There were three project milestones including the project end date. There were 50 feeding buffers. The initial project timescale was found to be late in terms of market delivery by some twelve months. This meant that the team had to go back and challenge all assumptions, task durations and the logical connections used. The final critical chain schedule including buffer came to just over twelve months of which if 30% of the project buffer was penetrated this would represent meeting the internal due date, and if 100% of the project buffer was penetrated it would still meet the external due date.

Once the team moved to execution they held regular meetings to check progress. These were done daily. Weekly they held progress/review meetings with senior managers where the penetration of the buffers was examined, with the percentage buffer consumed versus the remaining time on the critical chain being most important. They also kept a graph showing the cumulative penetration of the buffer with respect to critical chain completion to show how the project was performing.

Early lessons learned included the fact that completion of the critical chain tasks moves the whole project forward. The team also recognised the importance of watching the feeding buffers to ensure that no problems were being created there. There had to be clear definitions of tasks to ensure proper hand-over and the usual accolades given to those who multi-tasked had to be dropped entirely. They came to the recognition that not always do critical chain tasks get the right level of focus just because they are critical chain tasks and that not all estimations were based on the 50% probability of completion. In line with experience elsewhere they realised that all those involved had to have a clear understanding of the conceptual solution and the basic mechanics of it, plus an understanding of the TOC tools and their role in using them. The individual also had to understand the need for change when shifting from giving estimations from the usual elapsed times (90%) to the actual time (50%). The individual also had to recognise the importance of the buffer and the fact that finishing early would not attract penalty. Lastly, that if it is your turn to work – do it quickly. For the senior management there were also lessons to be gained. The first centred on the role of measurements. If buffers were penetrated the measurements had to direct those making decisions to the right decisions. It was accepted that it was the role of the senior managers to stop multi-tasking, and to do so through better prioritising of tasks in line with buffer penetration. They had to accept that it was no longer possible to punish the early finisher, or those who finish late. Perhaps most difficult of all was to recognise and fully accept that at some time people would not be working, but actually standing idle. This last was a major hurdle for some managers who place an inordinate amount of importance to the work ethic and the dominance of efficiencies as a realistic measure of performance.

So what of the results? In this case the drive hit the market early against the external due date by some five weeks. It was in fact the first drive of this type to arrive in the market. All competitors dropped out due to the technological challenges or by recognising that they were now going to be late to the market. Performance exceeded the previous 10 000 rpm product by some 30–35% and the company's customer satisfaction ratings were very high.

What were the final conclusions gained by this company? First was that people tried to beat the buffer, second that daily focus on the high buffer penetrated task is paramount. Next that it was essential to remove the barriers for the individual to operate effectively. They had to be helped and then they had to be rewarded. The location of the buffer at the end of the chain also eliminated the resting time, sometimes known as the student syndrome, and above all the most important lesson was to keep communicating and keep following up.

Other conclusions centred on the ability to develop a global metric for measuring time to market at all the design centres, another focused on the ability to identify product development constraints and a third of the provision of a clear line of sight for all members of staff. However, they also recognised that the culture is not changed with just one success. They are now working on creating more internal champions, in particular within the senior management group.

The final aspects to this case study involve the wider implications for the company as a whole. They see themselves as being the industry leaders in terms of cost, through economies of scale and process commonality, through vastly superior time to market with inherent flexibility to changes in the market, and the ability to be flexible in terms of scheduling right through the R&D and manufacturing to the market. They see the critical chain approach as delivering the goal of substantially reduced development cycle times and enhanced visibility on schedule adherence as well as better visibility, and resolution of resource conflict. This approach is now becoming part of the fabric of the company and will continue to do so for the foreseeable future. This is how they have moved from initial problems within the product development area, through the application of the TOC tools and management methodology to an enterprise wide focus.

Some of the key criteria for a successful implementation

What can be derived from this and the earlier case studies as the pathway for successful implementation? There are two distinct dimensions to the answer. The first lies in the project environment itself, and the second in the critical chain aspects. For any project to be successful it must meet a number of criteria, which require clarity. For example, in one company asking the project team members what the goal of the project was proved to be a quick way of inducing silence! They had not discussed, or been told, what the goal of the project was. Indeed, in one implementation workshop more than two days were given over to a debate concerning the definition of what a project was in their environment. The fact that it took two days to come up with a working definition which is even today, some two years later, still being discussed shows the increasing difficulties in obtaining clarity about an aspect to which everyone feels they know the answer.

So assuming that the goal of the project is known, how well are the features identified? How well have the specifications been detailed and understood? Have clear roles and responsibilities been defined prior to project start? What is the project structure? Has the work breakdown structure been defined? All of these questions demand answers and yet many times the projects being researched had not reached even this level of detail. It is clear that many companies have no formal structure for the process of project management, with many project managers able to work to their own rules about how they are going to deliver the project. This has proved to be a feature of quite a significant number of companies. The rationale behind this approach is that senior managers like to think this gives flexibility to their people, in fact, although the aim of flexibility is achieved, the aims of the projects are often put in jeopardy. It would appear that some senior managers allow democracy in place of leadership, usually because they have no idea, or process, to provide such leadership.

Once the networks were under construction other questions soon raised themselves. How sound were the logical connections between tasks? How was resource allocation to be achieved? Had the task durations been set according to the rules of critical chain? Had any risk assessment been carried out with respect to the content of the activities, and the timeliness of completion? Were the subcontractors involved in the activities where they would be expected to deliver?

In terms of the project definition and the goal of the overall project it was found necessary to provide, from the senior management team, a clear and unequivocal statement of what was expected as a deliverable. In one project meeting the list of deliverables expected for the whole project were listed. This list had been compiled by marketing in association with representatives of all the potential clients at the time. The project group then examined the list and decided to address only one market; one, which at the time was receiving a great deal of attention. Therefore only those features necessary for that market were allowed forward. From that reduced list a further reduction was achieved by removing all features that could not be done in the remaining timescales. From this much reduced list further reductions were made such that the key testing machine was not overloaded, bearing in mind that it had many other projects to work on. This left a much shorter list than that provided by marketing/client. On being asked whether they would inform both client and the marketing department the answer was 'no'. If they informed either the client or the marketing department they would be told to put the features taken out back in. This they could not allow as they knew there was no way they could meet such a demanding target. No attempt to try to deliver what the market wanted was made, no attempt to inform the market was made, they had a set of excuses they could use to explain away their non-performance so no problem there. This kind of scenario was quite common in the research and was a source of considerable frustration and stress for the managers and engineers concerned.

The leadership of the project function within this company had allowed itself to be sidetracked into a whole series of projects with no indication as to whether they could actually do them all. The assumption was that all projects had to be done, and there was no process to check the impact at any other point of the revenue chain, to the extent that the revenue chain did not figure in any decision-making process. This example is not an isolated one. It is an example of the kind of erroneous focus that has to be replaced with a clear understanding of what each project is trying to achieve, and the project function itself within the whole organisation.

Only by understanding who the primary and secondary customers are can real focus begin. These can be both internal and external. Next the ability to determine accurate and informative financial statements needs to be second nature. What are the costs of being late? What is the cost of a feature being compromised? What is the cost if we have to change focus during the project lifetime? Have we really defined the levels of risk and properly analysed and costed them?

Application of the thinking process tools of the TOC in the implementation

Having started to examine the nature of the enterprise focus, gained the buy-in of the people to the concept of the critical chain approach, and the management methodology that goes with it, the next step is to ask a simple question: 'what reservations do you have to implementing critical chain project management?'. As noted in chapter 2 and in Fig. 2.15 there are two distinct types of reservation, those that stop any one feature from being implemented, usually termed obstacles; and those which will cause, in combination with something which already exists, a negative outcome even worse than the original problem. This last reservation is called negative branch reservation or NBR within the TOC typology. Both of these types of reservation must be overcome if we are to be able to proceed to the final step in the improvement process, that of making it happen.

Dealing with reservations is all about the validation of the feature set of the solution, about ensuring that we know the steps to achieve each of the features, and the carrying out of a robust risk assessment for each feature prior to implementation, and before any major expenditure. Every implementation of an enterprise solution will have both reservations, there will always be people who can see obstacles in the way, and there will always be people who can see how things might go wrong. Rather than try to keep them quiet, the TOC approach is to encourage them as experience has taught us that if we ignore the reservations they will usually catch up with us in the end. Why not anticipate them and deal with them before they can cause any harm? Common obstacles exist for all implementations, just as there are similar ideas about where and how the implementation process can go wrong. They become part of the body of knowledge related to managing change. However, there are those that are unique to every implementation, which is why there is great focus in the TOC implementations on developing the skill of dealing with such reservations inside the organisation. This chapter describes some of the common reservations that occur almost every time, for a specific implementation you will need to have a TOC expert to help.

Network creation and validation

One of the common reservations relates to the ability to create and validate networks within projects of all sizes and scope. This has led to the development of programmes within implementations just to deal with this aspect.

The construction of networks was first described in chapter 3, but only at a technical level. What of the focus when looking at the whole enterprise? One question often asked is 'is the goal of the project plan the same as the goal of the project?'. The answer is 'no'. The project plan is designed to enable the project team to make decisions that help them achieve the objectives of the project. What about the role of the team at this time? Have the key players in the project been identified? Has a project leader been appointed, with clear lines of authority and responsibility? Have the resource managers been included in the project team? Have other contributors been identified, contacted and brought into the network creation process? What is the span of control with respect to outside contractors of all types? These questions may seem obvious but the research showed that often they were not addressed until a problem arose, by which time it was often already too late.

The level of task detail was a common issue at this juncture. How well defined were the features of the final product. This was true in all environments. What level of functionality was required and who set the specification? How were the various disciplines to be managed and integrated, and by whom? A common occurrence was the level of churn taking place. This in both the project content and the people involved. Once more issues of control and management came to the fore. How were these, often competing, factors to be resolved? Without a sound methodology for resolving such conflicts, experience had taught the players within the projects to play it safe, keep a low profile and not make too many waves!

In terms of the logic connections themselves a great deal of work was required. The project deliverables had to fully understood in terms of both the clients' wishes and the technical requirements. When and where did the external suppliers come in and what were the contractual conditions that applied? Were there links with other plans and how did they impact on the overall capability of both the primary plan being set up, and these other plans, all of which still had to be met?

The whole area of resource allocation gave considerable rise for concern. People were often found scheduling a single person some eight or nine months in advance with no real confidence that the person would still be in the company. Equally, given the fierce pressure to complete all projects in a short timescale, it was common to see multi-tasking scheduled in, even although the dangers were well understood.

Estimation versus execution

First discussed in chapter 3, this is perhaps one of the greatest sources of problems and reservations within project environments and occurs at a number of points along the path to the successful management of projects. During the implementation phase of a major critical chain (CC) project the problems of estimation, execution and the measurement system was made very clear. During a series of meetings to implement the CC approach the discussion centred, as it usually does, on the question of what times to use for the initial network creation. The rules of CC argue for the 50% confidence times, what I often call the actual times, whereas common practice uses the 90% confidence times, which I call the elapsed times. One is clearly longer than the other but even allowing for the knowledge that had already been gained in the company this conflict surfaced once more. The cloud is as shown in Fig. 5.1.

The logic of the cloud is as follows: In order to deliver results it is necessary to have a do-able plan. In order to have a do-able plan we must use critical chain planning methods. However, in order to deliver results we must also meet the committed due dates that have been dictated to us, and in order to achieve that we must use the traditional planning methods and practices. The project team felt that if they were

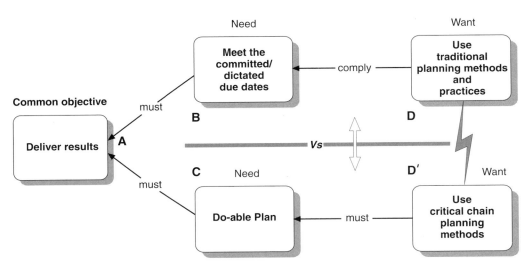

Fig. 5.1. A project network cloud

forced to use the traditional methods they would not have a do-able plan; however, if they used the CC methods they preferred they would not meet the dictated due dates, which is why they felt in a bind.

The logic behind using the CC methods rather than the traditional methods was clear; the issue was that the initial timescale for the project using the CC methods was too long. After careful analysis it became apparent that although they thought they were using CC methods they had in fact not done so. The times still contained between 60 and 75% confidence levels, in other words still far too high. They had not fully worked their way through the network validation process and there were still many occasions where the logic was faulty, or simply not challenged. They had to return to the initial project networks and re-examine the logical connections they had made, challenge the resourcing process, indeed challenge the whole way they constructed project networks, including the front end process of the project initiation, project objectives and timescales.

There was even a need to re-examine the risk assessments that had been done at the beginning and reassess the decisions made then. Once this had been done, as part of an iterative sequence of events, the timescales changed dramatically. The original timescale using CC including buffers now closely matched the requirements of the project director and the project sponsor. It was the process of achieving that result that was the most important. Achieving an effective, do-able critical chain schedule is not accomplished by just creating without challenge a network that appears to deliver the project objectives.

Once into the execution of the project plan the estimations are no longer of any value. They have been used to construct the plan and determine buffer sizes. Once the logical structure of the project plan is in place the next step is to execute the plan. Now the focus changes away from the estimations to the remaining durations of the tasks being done. Throughout the execution phase the only information of value about a task is the remaining duration, i.e. how long is it going to take before handover can occur? This is, of course, based on zero defects. It is not a point of failure to finish either early or late. If the task is finished early then there is a gain for the whole project, and the fact that the estimation might have been too large is of no importance whatsoever. Statistical fluctuation plays a role here, one time it might be larger than the estimation and another, smaller. If the task finishes late then there is penetration into the buffer, the reasons for the penetration is another function of the same statistical fluctuation. The only question that has to be asked is whether it was due to special cause or common cause, and that question can wait until the project is underway once more. This need to have a clear communication between the project team and the project management, and the resource managers, highlights the need for real trust and honesty between the various parties. The critical chain project management approach is an enterprise approach demanding a high level of trust between players.

Variation and how to manage it within projects

Now, as noted in the previous section, once the project is into execution it is the actual time to completion that is important, and the focus shifts to remaining duration. Having information about what has already been completed tells us nothing about when the task is going to finish, which is the most important piece of information. The estimated time is only a guide. The real time is the time it actually takes, and that is subject to variation.

The subject of variation requires almost a complete book in its own right. Variation lies at the core of estimation, it lies at the core of the conflict between estimation and execution. Variation is also a common source of reservations in many implementations. But variation is not easily described and requires careful definition before we can proceed. Deming (1986) defined variation in two dimensions, special cause and common cause. The common cause variation is that attributable to the system itself and special is simply that – special. The danger is to attribute variation as being special cause when it is in fact system noise, common cause, or to attribute variation as being of the system when it is in fact special. Either way, making both mistakes adds to the level of project variation.

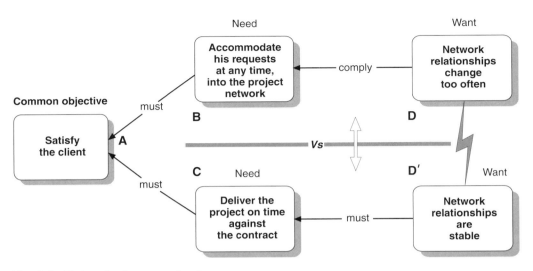

Fig. 5.2. Network changes cloud

A common failing in such sources as the *Project Management Body of Knowledge* is that it does not differentiate between common and special cause variation.

Within projects both types of variation will occur. The reasons for the existence of the variation are not our primary concern, the primary concern is to ensure that any form of variation has limited effect on the goal we wish to achieve. As negative variation will always take away time, it makes sense to use time as the buffer. Of course, should there be positive variation then the impact will be to allow the amount of time allocated to the buffer to increase, for which we see no real problem. Indeed, if it is possible to achieve positive variation which can be translated into additional buffer time, then the impact of this can be highly attractive. Deming argued strongly that no process can be measured properly unless it is under statistical control.

One example of externally derived variation came from one project environment where those tasked with the maintenance of the project networks were struggling with keeping the right focus. The problem lay in the fact that all too often the network changed as a result of changes due to marketing, or the client changing the content of the project. This led the group to develop the following cloud in Fig. 5.2.

The analysis of the content of the **B** box was the most interesting here. Bearing in mind that the content of the **B** box represents a conflict between constructs, that of the individual and that of the system, the pressure for the accommodation was coming from the marketing people who were keen to maintain a close relationship with the client as he represented a substantial amount of further business. To the network programmers, while they recognised the importance of keeping the client happy in terms of the requests, they also recognised that any change, especially those coming into the lifetime of the project, jeopardised the ability to meet the due date set within the contract. The cross-connection of the cloud was forcing them into potentially damaging rows in the project team meetings. What they came to recognise was that the system failed to give an indication as to whether instability in the network was the result of any change to the project deliverables. Although the individual construct in marketing was to always allow the request, they had no way of knowing whether it could be done or not, they just assumed it could. The rule was 'adopt any request without question'. What the system failed to provide was a clear signal as to approval of requests or not. The process of analysis of the requests was minimal. The first step in addressing this cloud was to address the differences between the two constructs contained in the **B** box. Once this was achieved, the next step was to develop the system in such a way that requests had to be subjected to scrutiny prior to any approval. Proper risk assessment had to be carried out, what–if scenarios developed and tested using the critical chain software, and finally the impact on the revenue chain assessed all prior to any approval, and all with the involvement of the client.

Multi-tasking – the devastating effect of this method of working

In a multi-project environment this aspect of operation is commonplace. Within single and multiple single environments it is less common but still has impact. This is perhaps the area of the greatest reservation. Many project people, while understanding the importance of removing the bad multi-tasking that goes on, have no belief that management will ever take the necessary steps to stamp it out. This is because multi-tasking is seen as the primary method of overcoming the requirement for a resource, person or some other manifestation of resource, to complete a number of project activities within the same time period.

When the research activity that forms much of the basis for this book was being carried out, this dominant theme in many project environments stood out. The allocation of key people to a number of concurrent activities or tasks within a project, or across projects was commonplace. In organisations where resource efficiency is the dominant measure, or at least one of them, this allocation method is highly attractive for resource managers seeking to ensure a good appraisal, promotion or, at the very least, the avoidance of penalty. By keeping all the people under his or her control busy they are rewarded by the system. The measure does not take note

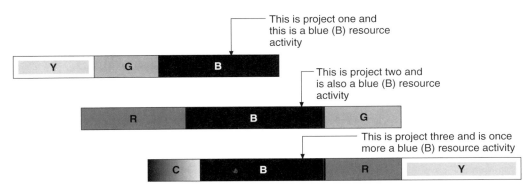

Fig. 5.3. Multi-tasking between projects – creating a resource bottleneck

of which activity is being done as long as activities are being done. In many cases all activities have an equal standing.

In Fig. 5.3 this can be shown in a simple fashion.

Here are three project plans with a key resource, blue (B), required by all three, and at the same time. This looks good, the resource is kept busy, if any one project falls behind, the resource can switch to another, and perhaps get ahead. Either way the resource, and it might be a piece of technology and not a person, has a high level of utilisation. If the green (G) resource activity of the first project runs late, the blue can always switch to the second providing the red (R) resource activity has finished. If there are delays in both project one and two the blue resource can start the third, assuming the cyan (C) resource activity has been completed. There might have to be some rescheduling of the other two projects when this happens, but the utilisation of the blue resource is OK, as is the utilisation of the other resources as they try to keep blue busy. Also there will be statistical fluctuations in the estimations of every activity and these too will have to be accommodated into the schedule, usually through a revision, or complete rescheduling of all projects.

To the project manager this does not seem quite so attractive. The resource is usually not available when required, which means delay, which for the project manager is not good news. Often in implementations this conflict arises, and can be seen in Fig. 5.4. The individual resources might be doing well, but the projects they are allocated against are most certainly not in good shape. They are experiencing quite considerable fluctuation in terms of meeting due date, keeping to budget, time for the rescheduling exercises and so on.

Note the power of the cross-connection in this cloud. **D**, the lack of prioritisation, often results in **C** being jeopardised. Many projects do not meet their objectives due to the negative aspects of multi-tasking. The delays caused as a result often lead to specifications being compromised

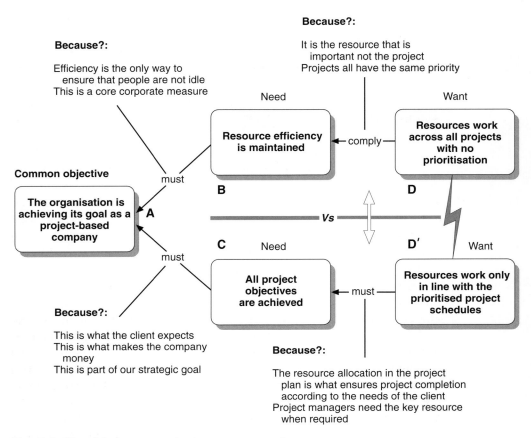

Because?:

It is the resource that is
important not the project
Projects all have the same priority

Because?:

Efficiency is the only way to
ensure that people are not idle
This is a core corporate measure

Need

Want

**Resource efficiency
is maintained**

←comply—

**Resources work
across all projects
with no
prioritisation**

B

D

Common objective

must

**The organisation is
achieving its goal as a
project-based
company**

A

━━━ Vs ━━━

C

Need

D′

Want

must

**All project
objectives
are achieved**

← must —

**Resources work only
in line with the
prioritised project
schedules**

Because?:

This is what the client expects
This is what makes the company
money
This is part of our strategic goal

Because?:

The resource allocation in the project
plan is what ensures project completion
according to the needs of the client
Project managers need the key resource
when required

Fig. 5.4. Cloud between project managers and resource managers

in order to finish on time, or an increase in the budget, or both. It is the project manager who has to field the awkward questions, not the resource manager, after all the resource manager seems to have met his measurement of keeping his/her people busy most of the time. At the same time **D′** attacks the heart of the corporate efficiency measures. Under critical chain project management (CCPM) there will be excess capacity revealed. Indeed all resources require an element of spare capacity to act as protection. Within CCPM the argument is that by working smarter, through the eradication of multi-tasking, the actual duration of all activities comes down and the overall performance of the organisation is enhanced. This represents a real enterprise focus. No longer is the local optimisation of a resource of any value, only the optimisation of the critical chain for each project, and the strategic resource in multi-project environments.

This is a fundamental shift of paradigms for project-based companies. In the exercises used to introduce the multi-project approach to CCPM, Tony Rizzo devised the bead game. Using coloured glass beads and a simple project structure, Rizzo was able to show how to gain between 50 and 60% more productivity from the same number of people, doing exactly the same tasks. We ourselves have run this exercise many times, well over a hundred, and the results are always in line with expectation. When we have asked the managers present what they might be able to do with such improvement in productivity, they always reply by stating that this would enable them to have far greater confidence in meeting the existing targets in terms of the number of projects, and probably take on more projects without overloading the system.

The impact multi-tasking has is immense. It seems to have been accepted as a necessary evil. The reason for this is simple. It is the perfect example of a rule, a procedure, that has been developed in order to get around a limitation. However, CCPM and the associated software represent a technology that removes the limitation: how to prioritise so many projects, so many activities within each project, and schedule effectively a strategic resource. Only an enterprise resource planning (ERP) approach has the capability to do this. If, however, the procedures used to work the limitation in the past are retained, then implementing an ERP project management approach will only result in substantially more problems. This is an excellent example of the NBR described earlier, a really negative outcome of implementing what should be an outstanding solution.

If matters were to be left in this position the resource managers, who typically have more power in most project-based industries, would over-rule the project manager until the situation was really out of hand. The new solution of CCPM would be seen as the spark for this particular fire, and before long the solution itself would be disowned or discredited, or both. Note that the original limitation remains, the procedures remain, the measures remain, which means that the organisation just ceased to improve. Effectively local optima has triumphed over global optima. All of what Deming, Goldratt and I have been arguing for would appear lost. The source of this analysis lies at the heart of every ERP implementation, the need to challenge and to change, if necessary, the rules, procedures and measurements every time a new technology is proposed and implemented. This analysis starts in the pre-sale activity, continues through all aspects of the sell cycle, and on into the implementation phase, and post implementation support. The management methodology that lies at the heart of the CCPM enterprise solution, the TOC/TP, is the only process I know of that can deliver the answers to these questions.

Overcoming the problem of multi-tasking and key resources

Dealing with the devastating impact of multi-tasking is a key requirement for anyone managing projects where this has become an issue. The

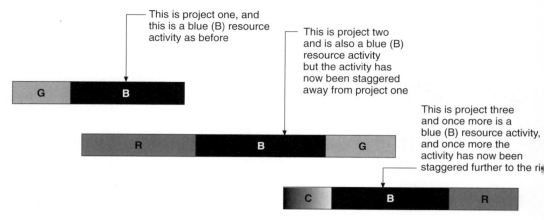

Fig. 5.5. Staggering the strategic resource between projects

answer is simple conceptually but often appears difficult in practice. The method employs staggering the strategic resource between all projects, and buffering accordingly. This is shown in Fig. 5.5.

Note that the blue (B) resource activities have been shifted in time along the x axis. The intention is that the blue resource will start with the activity of the first project and complete it as quickly as possible. The gap between the completion of the first project and the second project in terms of the blue resource is simply to ensure that should there be any delay in project one, project two is not affected. If there was no time buffer between the two blue resource activities, and project one was a little late, this would have the effect of not only consuming some of the protection of project one, but also of project two as well. The individual projects must be properly protected from the impact of variation within the project, but it would be very difficult to try and protect every subsequent project as well, hence the need for decoupling the projects, one from another. Other buffers will also be required to ensure that the strategic resource is not kept waiting. Therefore those activities that feed to the strategic resource, for example, the red (R) resource in project two above, should finish a little before, just in case there is a problem with that particular activity, or the blue resource is able to complete the activity in project one earlier than expected. We do not want the blue resource to be waiting for the red to complete as we will have missed the opportunity to move the whole company forward. This is shown in Fig. 5.6.

The gap that has appeared between the blue resource activity and the preceding activity is a buffer to protect the operation of the blue resource. The combination of staggering, and the additional buffering, ensures that each project is able to deliver on or ahead of time, the blue resource is

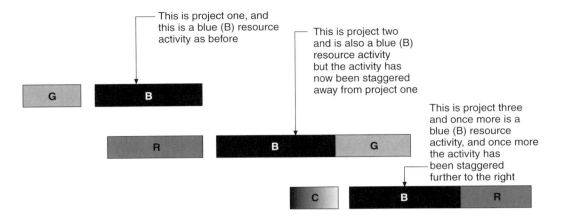

never overloaded and the whole organisation gains. For this activity within multi-project environments it is essential to use appropriate software, and a product such as that from Speed to Market entitled *Concerto* is but one example.

The real issue is not software, however, but the management thinking that accompanies the software. The constraint of each project is the critical chain. The constraint of the whole organisation is the strategic resource. It carries the highest priority, after that has been properly subordinated to, it is possible to subordinate to the constraint of each individual project, the critical chain. By so doing the limitation is removed, multi-tasking is no longer needed, and projects deliver what they are supposed to deliver. Even if this is the only feature of CCPM implemented the organisation will be in a far better shape than before, and have much higher confidence levels of meeting aggressive project targets.

Buffers and buffer management – the method of control

Much has already been said about buffers, their location, their size, their role in protecting the projects, and the strategic resource in the delivery of the project. Not much has been said about managing buffers. This is perhaps one of the most interesting reservations raised during an implementation. There are two distinct aspects to buffers, the technical and the managerial. In terms of the technical almost any critical chain software will accomplish this task without a problem. All the project planner has to do is to determine the buffer length as a proportion of the chain, feeding or critical, and the software will do the rest. There are times when variations to that theme will be accommodated but once more there is typically no problem with that. The real issues centring on buffers is how to manage them, what to do when there is a penetration, when not to react and when to implement a strategy for bringing the project back under control.

The use of buffers is therefore the key dimension in both the control over the project itself and allowing for the ability to make decisions, which deliver the goal of the project. Managing the buffers properly is also the key element in the successful management of the enterprise focus. It is the buffers of both the individual projects, and in multi-project environments the strategic resource that ensures the proper focus is maintained. Here the reservation is usually one of being an obstacle: just how do we manage buffers? What are the issues that need to be addressed in order to successfully manage buffers, from top–down? This often entails a substantial amount of training in the role of the buffers, and great care in the implementation of the relevant software to ensure that the buffer reports so generated give good data upon which to make proper decisions.

Buffers are measured in time. As already explained there are three primary types of buffer in all project environments: feeding buffer, resource buffer and the project buffer. Within multi-project environments there are other buffers: the drum buffer and the strategic resource buffer. The buffers are for the protection of the project, for both the customer and ourselves. They are vital to the well-being of the project. In essence they should be sized according to the length of the path feeding them. However, there have been examples of buffers being far too small and yet still performing as they should. The very first project within the construction industry operated with a one-month buffer for an 18-month project timescale. This was far too short yet the project completed on time. The answer, of course, is that the buffer size is not fixed but dynamic. As a task finishes earlier than the duration set for it there is a positive variation in the buffer size. Equally when a task overruns its allotted duration the buffer is consumed. This variation in the size of the buffer once we are in execution mode means that if we feel that the estimated task times are still containing elements of elapsed time, once the task is actually being performed it will take only the time necessary to complete and therefore the buffer will either grow or decrease.

Buffers being time means that it is necessary to have some kind of measurement to give us information about the status of the buffer. Just what is the penetration of the buffer in terms of both the size of the buffer itself and the state of the project?

Let us examine the simple project from chapter 3 (Fig. 5.7).

Once we are in execution mode the activities are completed according to the rules of critical chain. The project managers, the team leaders and

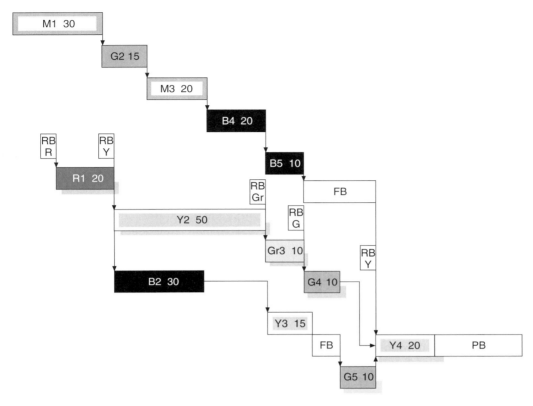

Fig. 5.7. A complete project network including buffers

the senior management group are all focusing on the same set of criteria, remaining durations of activities and penetration of buffers. In order to assist with the focusing capability, the overall length of the buffer, measured in time, is split into three zones. The first time zone, shown in green, is the furthest away from the due date. There will always be some form of penetration to this zone throughout the project lifetime. If there is no penetration then it is fair to say that the overall buffer size is far too big. The green zone is usually described as the 'watch' zone as there should be no reaction to a penetration at this level. Watch what is happening certainly, but do not react. The next zone, shown as yellow, is the zone where the level of penetration is attracting more interest. Now the remaining level of buffer has to protect the remaining level of the chain feeding the buffer. The balance needs to be examined. If the completion date is close, if the remaining activities are few, with little or no real time involved, then there is probably nothing to worry about. If, however, the remaining amount of the feeding chain is high, and the level of remaining buffer deemed insufficient to give the proper level of

117

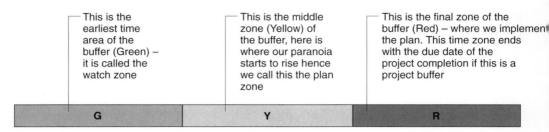

Fig. 5.8. *The basic structure of the buffer*

protection then a plan should be developed for implementation to remedy the situation. The plan is not yet implemented. The system might still be able to retrieve the situation without any input from the project manager. This is allowing common cause variation to sort itself out and not respond to all variations as if they were special cause. Of course there are times when the cause is special cause, that the plan must be implemented and properly controlled to ensure that the buffer continues to do its work. Now the management of the buffer incorporates a degree of management thinking and analysis, the buffer reports have to be examined and properly understood before changes are made. Let us examine this in more depth.

Here we have a simple buffer (Fig. 5.8), the three zones are shown and what they represent is also shown. Note that for this exercise they are all of the same time duration. This might not always be the case. Once the project is underway the remaining durations of the tasks along the chain are submitted to the centre and the project database upgraded accordingly. If everything is running to time then there is no penetration of the project buffer. If, however, tasks are being completed early then the overall time of the buffer is being extended.

Remember the distinction between the two primary types of variation and the mistakes that are often made. Mistake one is reacting to every outcome, every perturbation in the system, as if it came from a special cause when actually it came from common causes of variation. Mistake two is to treat every outcome, every perturbation in the system, as if it came from common causes of variation when in fact it was due to special cause. This has important connotations for how we carry out risk assessment in the area between zones three and two. If the variation is a result of the system, then part of our analysis should reflect this and our developed plan should also reflect this. If the variation is due to special cause then the level of paranoia will be different, and so will the plan developed to address the situation.

The importance of this debate is the understanding about processes being either in statistical control or not. Joiner (1994) states that

Processes with only common causes are stable, or predictable; you might hear them referred to as in statistical control. Processes with special causes are unstable, unpredictable and not in statistical control.

Joiner describes the differences as far as management is concerned as

The differences between common cause and special cause variation thus require us to use different managerial approaches to deal with each if we are to be effective. With unstable processes, ones with special causes, we are interested in finding out what was different when special cause appeared. With a stable process, one with only common causes, the answers are often more subtle. One of the biggest revelations in dealing with variation is that most problems arise from common causes. Yet paradoxically it still pays to work on special causes first.

So, given the importance of understanding the two distinct types of variation, what happens in reality with the buffers? Consider the first scenario, when the tasks within the chain are all finishing on time. The buffer reports for the main project buffer will show no penetration, the feeding buffers will also show if there are any problems with the feeding paths. If all goes well the project will actually finish at the start of the project buffer. In other words the customer is able to take delivery a buffer time ahead of the original due date. Of course, if the client is unable to take delivery, cannot use what is being produced until the original due date, then all that happens is that the handover is delayed. The resources, however, have already been released to work on other projects or help out on existing projects.

What happens when there is a delay? The first question is how long is the delay and what impact is it having on the buffer? If the delay is short, only into zone three, and the remaining time also short then there is probably no need for concern. If the delay is short but the remaining time is still long there might be a need to watch this with more concern, but no more than that. If the delay is going to consume a substantial part of zone three with a great deal of the project to run then it might be prudent to think about what remedial action could be taken and to plan for it. Not to implement the plan, just develop one in readiness. Once the delay extends into zone one there is only one option – implement the plan immediately. Of course it is possible that other events will help. Early handovers add to the overall buffer size and therefore can give more protection than was originally envisaged. Buffers are dynamic, their size will fluctuate in line with the finishes being reported. Early finishes add to the project buffer, late ones diminish it. Therefore the buffer management reports also have to take notice of these fluctuations. Often quite considerable variation in one activity can be safely accommodated without too much trouble simply because of the early handovers of previous activities. This is why such handovers are so important. It is of no value watching people handover always on time if they could have handed over early. That suggests there is still work to be done in the measurement systems.

The behaviour patterns of the project staff are still not in line with the conceptual approach of critical chain. The same applies to managing the buffer through the budget. Critical chain attempts to deliver the project to budget by focusing on the constraint on the project itself, the critical chain. There is still a danger, however, that project and resource managers try to manage the critical chain through the use of the budget. Joiner (1994) highlights some of the dangers of doing this when he writes

> The disadvantages of managing to budget are numerous.
>
> 1. promotes short-term business decisions; optimises the piece, the cost of the individual project, at the expense of the whole
> 2. customers, quality, and perhaps even safety receive short shrift
> 3. stifles creativity (except in encouraging people to find creative ways to meet the figures)
> 4. people spend whether they need to or not
> 5. leads to tampering.

This is a real problem within some project environments where budgetary pressures force decisions to be taken which do not assist the project, or the client, but give a supposed short-term benefit to the company.

Conclusions

This chapter has addressed the role and nature of enterprise focus within project-based companies. It has covered some of the common reservations raised when the implementation of the enterprise approach is being discussed and carried out. Keeping a focus on the whole of the revenue chain is not an easy task, it requires people to understand the revenue chain, be able to describe it and understand their contribution to it. The enterprise focus demands alignment of measures, of rules, of procedures in order to function and deliver the real benefits that the approach contains. The use of technology through the use of enterprise resource planning systems lies at the heart of such an approach, but the lesson is that it is not technology that delivers the benefits, it is people who can manage the technology with the right level of focus and leverage who will be successful. One dimension is already clear from the majority of the work being done to date in such industries, that change plays a fundamental role in the success or otherwise of the investment. It is of no value whatsoever in carrying out an analysis of your company – an enterprise analysis, developing a solution, complete with the appropriate technology – if the change within the organisation cannot be successfully implemented. No positive result will accrue, the investment will have been wasted, and the organisation put at risk. The next chapter examines the change process related to problems and organisational issues, change models, and some of the key issues that must be borne in mind if the organisation is to benefit.

CHAPTER SIX

The strategic importance of managing change

Introduction

This chapter describes the models of change and the need to ensure that the proper attention is given to change issues. If the whole area of change management is ignored there can only be one result – disaster. There are a number of fundamental changes that must take place within any organisation seeking to adopt, and gain benefit from, an enterprise focus. There is the shift to the critical chain approach for project management. There is the need to adopt a sound management methodology such as that contained within the TOC thinking processes and taught through the enterprise analysis programme. There is the need to change some of the measures used in support of decision making such that there is proper alignment between decisions and overall direction. Above all, there is the need to change the behaviour of many people within the organisation. This last aspect is no easy task and should never be underestimated.

The role of change and change management

This book has followed a simple model, that all projects are designed to deal with a problem. The starting point is the recognition that a problem exists. This leads to a careful analysis of the problem and from that to develop a solution. Once the solution has been determined it is important to implement it properly. This usually involves change, from a small minor adjustment to considerable upheaval throughout the whole organisation. To be successful this change must be properly managed. This cycle of problem–solution–implementation is also linked to the ability of the organisation to learn, from both failure and success. In the world of project management the first question that must be asked is 'is there a problem with the way we do things?'.

Newman (1995) defines problems in four ways.

A performance deviation is where something odd or unexpected has occurred. A matter of difference is the gap between where we are, and where we want to go.

He then adds two further aspects.

> An open problem is one without a correct solution and a closed problem is one, which can be precisely defined, and has clear parameters and a correct solution.

In the sense of the focus of this book, the types of problems being discussed are often seen to be open. There is no clear solution. Indeed where the deeper questions raised by the study are concerned, the individual is almost always convinced that there is no solution. He is also convinced that the gap is too large, that there is no process that can bridge the gap. Newman recognises the impact that mindset can have on this element of problem solving. He also recognises the importance of learning through problem solving.

For Newman the first stage is to define the problem. He then discusses a number of techniques to do just that. Analytical tools, such as fishbone diagrams, multiple cause diagrams and force field analysis, are described by Newman as viable tools for achieving the level of understanding required. Once the data has been analysed, he suggests using such techniques as brainstorming for the generation of solutions. Once a number of potential solutions have been identified it is then necessary to choose one. When this task has been accomplished it is then necessary to implement the solution. Within the context of this book the tools used were those of the TOC/TP and described earlier.

VanGundy (1988) has defined a problem as

> ... any situation in which a gap is perceived to exist between what is and what should be. If an actual and a desired state are viewed as identical, then no problem exists.

VanGundy summarises the preconditions he feels are necessary in order to begin the problem-solving process as being

> 1. the existence of a gap between what is and what it should be
> 2. an awareness that a gap exists
> 3. the motivation to decrease the gap
> 4. an ability to measure the size of the gap
> 5. the abilities and resources required to close the gap.

VanGundy describes what is almost the definitive sequence of events in problem solving. Starting with problem analysis and redefinition, through idea generation to idea evaluation and selection and ending in implementation, this sequence is deemed to be the norm. He and Newman describe a whole range of potential methods that can be used, but they are all stand alone. The work carried out as part of the development of this book highlighted a need for a different approach. This suggested that a more systematic method would be of greater value. One such approach is that proposed and outlined by Checkland (1981) and Checkland and Scholes (1990) called *soft systems methodology*, which will be discussed later.

Newell and Simon (1972) suggested that human cognition be based on the ability of the individual to process information. This includes the ability to store and retrieve information from long-term memory, and at the same time have the capacity to handle information in short-term memory. This led to the recognition of two key phases in problem solving, identifying the problem space, and being able to use some form of means/ ends analysis for solutions. The first element is part of the intuition and knowledge of the individual, being able to recognise problems, or the patterns that determine problems, based on the previous experience of Newell. The second element is about the ability to determine, or select, actions which take the individual closer to their goal. These actions are then implemented.

Of course, for many situations, more than one type of action is possible and it is not always feasible to either remember each one, or to determine the impact it might have. There are also occasions when the correct path is to take actions that appear to move away from the goal, but which are, in fact, necessary in order to achieve the goal. These actions can often be in conflict with accepted practice.

Newell and Simon (1972) indicated several ways in which the second element could be successfully implemented. These include

- specifying and attaining sub-goals
- working backwards from the goal to the solution
- using old analogous solutions on the current problems
- using diagrams of various sorts to delineate the problem space.

The role of intuition is recognised as vital if the initial analysis is to have any merit. The ability to properly define the problem space and thus the core problem is the first step in determining the solution.

Checkland (1981) sets his problem-solving methodology into a scientific context. Drawing on the research developed in the area of general systems theory, Checkland considers the importance of placing problem solving firmly in the scientific domain. For Checkland the importance of applying the scientific approach is in the determination of explanations which, for him, '... requires the elucidation of chains of causes and effects, and testable prediction'. For Checkland, logical analysis is a vital part of understanding what is happening within the system. He does sound one note of caution however. He writes

Scientifically acquired and tested knowledge is not knowledge of reality, it is knowledge of the best description of reality that we have at that moment in time.

Checkland argues strongly that science is an enquiring or learning system. He goes on to write

Science is a way of acquiring publicly testable knowledge of the world, it is characterised by the application of rational thinking to experience, such as is derived from observation and from deliberately designed experiments.

This application of the scientific method is a key feature of the approach adopted by Checkland. In terms of hypotheses, he argues that '...a hypothesis refuted is a more valuable experimental result than one in which the hypothesis survives the test'. Given this background in the scientific method Checkland developed his approach to problem solving as comprising the following steps.

1. the problem situation: unstructured
2. the problem situation: expressed
3. root definitions of the relevant systems
4. conceptual models comprising both a formal system concept and other systems thinking
5. a comparison between the conceptual model and the problem situation
6. the development of feasible, and/or desirable changes
7. the action(s) to improve the problem situation.

What is clear from the work of Checkland (1981), Checkland and Scholes (1990) and Katz and Kahn (1978) is that the systems approach differs greatly from that of Newman (1995) and VanGundy (1988). The systems approach takes note of the causality that exists within organisations. This is very much in line with the TOC/TP approach that demands the viewing of the organisation as a series of links in a chain whereas that of Newman and VanGundy makes no such assumptions. The traditional approach considers each link in the organisation to be separate and that the improvement of any one will lead to an overall improvement. The systems approach, with the focus on the links being part of a chain, notes the importance of the interdependence of the links and argues that this must be taken into consideration when trying to deal with problems.

As already described in some detail, Goldratt (1997) argues that as the organisation must be viewed as a series of links in a chain, then the efforts in terms of solving problems must be focused on the weakest link in the chain. Equally, any actions that are taken as a result of the problem will inevitably have impact elsewhere, due to the linkages. Therefore the importance of managing the change process assumes a greater degree of significance. If the impact were in only one area with little or no impact in any other area, then the process of change would be primarily in that one area. If there are linkages, the change process will impact a far wider environment than before and create new problems, in particular the need to resolve potential conflict. Recognition of this impact means that anyone attempting to use the TOC/TP approach to problem solving must be aware of the systemic nature of the process. This requires a careful understanding of many of the key issues raised by managing a change process.

Managing change

Brooks (1980) described the elements of organisational change as shown in Fig. 6.1. From this Brooks derives five major areas which management must take into account if it is to aid rather than hinder organisational change. Brooks considers

> *The model provides a conceptual framework which focuses on the key variables influencing the success or otherwise of management change initiatives.*

These five are the aims and objectives of management, the technology being applied, the people involved, the current structure, and the range of control in the environment.

Of course when considering the forces that act on organisations, the structure of that organisation has important connotations for the way in

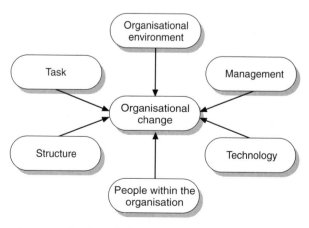

Fig. 6.1. Brooks organisational change model

which it responds. It is implied that if the organisation and the environment in which it exists is about to change, then it is equally likely to require a change within itself in terms of structure and culture if it is to remain responsive and adaptive. There have been many studies, which show that there are many effective organisations operating in stable environments or with stable technologies that are characterised by rigid structures with power concentrated at the top, and clearly defined roles at lower levels. It is equally true to say that where the environment is rapidly changing then the effective organisation is characterised by less reliance on formality and greater reliance on interdependence of unit operation. This is coupled with greater emphasis on joint planning and problem solving with greater responsibility and authority placed at lower levels. (Burns and Stalker 1966; Emery and Trist 1965; Lawrence and Lorsch 1967; Pettigrew and Whipp 1996)

Thus, if an organisation is in the process of change, fixed rules and procedures will rapidly become outdated and severely hinder the process. As change often involves the unexpected, with unforeseen influences, a joint participative approach is to be preferred, as is the case with the TOC/TP methodology. Of course some organisations still feel that change will not affect them. This view is dismissed by Hersey and Blanchard (1972) who take the view that such is the nature of a dynamic society that the question of change has shifted from whether it will happen to one of when. They state

> ... how do managers cope with the inevitable barrage of changes which confront them daily in attempting to keep their organisations viable and current. While change is a fact of life, effective managers ... can no longer be content to let change occur as it will, they must be able to develop strategies to plan, direct and control change.

The risks associated with change

The fact that change has become commonplace involves the organisation in risk-taking. Moore and Gergen (1985) place risk-taking as

> ... a crucial element in change, transition and entrepreneurship. In turn, fear of risks is a key factor in resistance to change both for managers who need to decide whether or not to initiate change and for employees required to adapt for change.

Thus risk has two primary elements, the risk to the organisation and the risk to the individual. Both of these are recognised by Moore and Gergen who then outline four key structural/cultural factors that influence risk-taking. These can be summarised as listed below.

127

- Organisational expectations where the managers need to clarify what changes need to occur, why they are necessary and what is expected as a result of those changes.
- Reward systems, whether formal or informal.
- Support systems, which apply to the entire workforce.
- Available resources to allow the risk taker to discover a working system.

In order to achieve change Moore and Gergen consider that organisations need moderately high-risk takers and it is this risk that requires careful consideration. They conclude by saying

> *Asking people to change is asking them to innovate: to try new tasks, skills and work methods at all levels to make the change work well for themselves and the organisation.*

The challenges of change

Risk is not the only challenge facing managers when change is required. Leonard-Barton and Kraus (1985) identify a number of key challenges which include the dual role within the company of those involved with the task of change, the variety of internal markets to be served, legitimate resistance to change, the right degree of promotion, the choice of the implementation site, where appropriate and the need for one person to take responsibility. In consideration of the dual role they note that

> *Those who manage technological change must often serve as both technical developers and implementers. As a rule, one organisation develops the technology and then hands it off to users, who are less technically skilled but quite knowledgeable about their own areas of application.*

Although their focus is that of technological change, what they argue here is also true of almost any change process and is certainly true of the TOC/TP change process. This imposes a responsibility for the implementer to design the changeover in such a way that it is almost invisible. When the people are transferring the TOC/TP knowledge contained within the CCPM approach into their own environment it is vital to ensure that the others within the organisation are ready to accept the new approach without question. There is also a major question here for the TOC/TP developer and educator. If they are to be successful then they must also be able to integrate the needs of both parties. For Leonard-Barton and Kraus (1985) the way forward is through a marketing approach. They argue that the

> *Adoption of a marketing perspective encourages implementation managers to seek to use involvement in the*
>
> 1. *early identification and enhancement of the fit between a product and user needs*
> 2. *preparation of the user organisation to receive the innovation*
> 3. *shifting of 'ownership' of the innovation to users.*

The time given to achieving the buy-in of the other members of the organisation is felt to be a prime factor in the success, or otherwise, of the process.

Managing the introduction of change

Wooldridge (1982) notes the concerns of managers facing the introduction of change by saying

> *...the prospect of introducing technological change has brought about increasing despair amongst line managers. They cannot believe it when faced with outright opposition to change from employees and their unions, even when that change is blatantly vital to the survival of their organisation.*

One of the major problems facing anyone concerned with this is that the reality of the situation is both complex and subtle. Again Leonard-Barton and Kraus (1985), who argue that the marketing approach will assist in the change process, have found that many implementations fail because 'someone underestimated the scope of importance of such preparation'. The idea that the technical superiority of the innovation will guarantee acceptance and that pouring abundant resources into the purchase and development of the technology at the expense of the implementation process will provide success is scorned by Leonard-Barton and Kraus. They propose 'not only heavy investment by developers early in the project but also a sustained level of investment in the resources of user organisations'.

Without proper implementation there is no improvement process and if the involvement of all the people concerned is not achieved then the likely outcome is not what might be expected. The whole point of the third question 'how to effect the change?' in the three questions of TOC/TP is to fully prepare the people for the change and indeed to involve them in that process. This is done through the surfacing of the obstacles that stand in the way of success and the raising of reservations, which refer to the possible negative outcomes of implementing the proposed solution. The notion of the marketing approach also implies that there are multiple markets within the organisation which require to be addressed. At each level of the company there needs to be a planned strategy, which applies at that level. Managers on one level will require a different response to managers or users at a different level. Leonard-Barton and Kraus believe that

> Top management and ultimate users have to buy into the innovation to make it succeed but marketing an idea to these two groups requires very different approaches.... We believe this executive must view the new technology from the perspective of each group and plan an approach accordingly.

This buying-in to the proposed innovation leads to the concept of 'ownership' of the proposed change. Although the exact meaning will vary depending on the size and nature of the change project, the term implies the full involvement of all interested parties. Of course one set of difficulties not expressed here is when there is full buy-in from the people but not from the innovator, for reasons which will be examined later in this study. Of course any change of this nature involves a transfer from one set of technology to another. Therefore the identification of those who will influence the workforce is paramount. The opinion leaders within the organisation play a vital role at this stage. Equally important is the realisation that the opinion leaders may not be the actual managers within the departments, but others who, although they do not have the functional leadership, may have the *de facto* leadership.

The change agent

Whoever is responsible for the change, they all share one thing in common, they are agents for change. Atkinson (1985) notes the importance of the change agent when he writes that

> The key to change is recognising that a need to take action is an important aspect of the change process, and that a 'change agent' or 'catalyst' is imperative to the successful implementation of new technologies.

It is the nature of this person that is crucial to the success or otherwise of the implementation. Atkinson suggests the following description of the person who can deliver such change

> The individual, the catalyst who makes things happen, is central to effective implementation strategies. Whoever occupies this role must possess the requisite attitudes, skills, knowledge and experience to develop an objective overview of the problem, and create decisions which work in the long term. This 'facilitator' must also be able to harmonise enthusiasm, grasp opportunities, explore the

sources of resistance and change attitudes in order to promote a healthy and effective organisation ... change cannot create itself. It must be welcomed as an opportunity for those who work in the organisation to take some responsibility, and help prepare for, and create their own future.

Hersey and Blanchard (1988) picked up the theme of motivation and leadership, two key aspects of a successful change agent, when they described what makes a good leader. Through the development of situational leadership they sought to provide a 'common language to help solve performance problems'. They go on to suggest that their approach can be used to 'diagnose leadership problems, adapt behaviour to solve these problems and to communicate solutions'.

In their analysis of what they call a 'real change leader' (RCL), Katzenbach (1995) and his team consider the key elements or attributes of a successful change manager. He defines such major changes as

... those situations in which corporate performance requires most people throughout the organisation to learn new behaviours and skills. These new skills must add up to a competitive advantage for the enterprise, allowing it to produce better and better performance in shorter and shorter time frames.

He also describes the common characteristics of the RCL as

1. *commitment to a better way*
2. *courage to challenge existing power bases and norms*
3. *personal initiative to go beyond defined boundaries*
4. *motivation of themselves and others*
5. *caring about how people are treated and enabled to perform*
6. *staying undercover*
7. *a sense of humor about themselves and their situations.*

-This leads to their definition of the term 'real change leader' as

Individuals who lead initiatives that influence dozens to hundreds of others to perform differently – and better – by applying multiple leadership and change approaches.

Throughout his study Katzenbach uses cases to describe what kind of effect RCLs can have, in particular they go after '... specific performance improvements based on building new skills and attitudes, and getting commitment from all hands.

The close relationship between what Katzenbach calls an RCL and a TOC/TP practitioner is not surprising. Both have a focus rooted in the market, both are working to a vision of what can be, and setting out to achieve it. Words that are part of the language of a TOC enterprise analyst, known within the TOC community as a Jonah, such as enthusiasm, commitment and trust are also part of the RCL vocabulary. Katzenbach recognises the importance of vision in the process of change. He writes

> *You have to start out with a vision that isn't well articulated. You have to sense that it can be grey, it can be murky, but it should have at least the attributes of things you want. Then I think it is helpful to talk long and hard to a lot of people.*

The next stage in the process is to determine whether the agent of change should come from inside the organisation or outside. In one sense, given the nature of this research and the role of the researcher, the answer to this question is both inside and out. The primary force for change is with the people attending the various programmes; however, the researcher in his role of educator is also influential in the change process. Atkinson (1985) considers both.

> *Clearly there are many advantages associated with using personnel within the structure to bring about change. Organisational knowledge, relating to structure, organisational culture, departmental and work groups, managerial responsibility etc., is an important asset, which the internal change agent possesses. Unfortunately there are disadvantages regarding subjectivity, bias and dependence for future career and promotion prospects. These negative factors all tend to suggest that the internal agent for change is not sufficiently detached from the situation to perform his function with discretion.*

When considering the position of the agent being external Atkinson argues that their advantages lie in areas such as

> *. . . skill attainment, experience of change programmes in different cultures and structures, objectivity etc. All this helps the external practitioner to take a detached and professional view. Unfortunately, external consultants take a great deal of time to become acquainted with organisational philosophy, policy and practice.*

Caruth (1974) in his examination of the systems analyst as a change agent noted that there are

> *. . . four areas of major concern to the systems analyst*
> 1. *basic human motivation*
> 2. *why people tend to resist change*
> 3. *the ways in which people resist change in the workplace*
> 4. *how to overcome resistance to change.*

Of the first area Caruth considers that this is a key area on which to focus. It is necessary for long-term improvement and change to reinforce the motivation of the workforce and that the 'greatest opportunities for motivation lie in the areas of egotistic and self-fulfilment needs'. It is often due to the lack of opportunities for most employees to use the creative side of their personality in their work and to a lack of recognition and appreciation that they become stagnant in their approach. Again Caruth makes the point that

> *There are areas in which management must concentrate its moti-*
> *vational efforts. These are the areas which the systems analyst*
> *should utilise in his efforts to bring about change with a minimum*
> *of disruption and resistance.*

Atkinson (1985) considers that the best approach centres on what he calls the team approach. This he describes as the

> *... coming together and grouping of external specialists who possess*
> *expertise, skill and experience, coupled with the organisational*
> *strengths of the internal practitioner, helps bind the partners of*
> *change. The team approach develops a 'synergistic' learning climate*
> *where the experience, knowledge and creation of ideas can be*
> *maximised.*

Leonard-Barton and Kraus (1985) also considered the formation of the implementation team and argued that it should include

> 1. *a sponsor, usually a fairly high-level person who makes sure that*
>   *the project receives financial and manpower resources and who*
>   *is wise about the politics of the organisation*
> 2. *a champion, who is a salesperson, diplomat and problem solver*
>   *for the innovation*
> 3. *a project manager, who oversees administrative details*
> 4. *an integrator who manages conflicting priorities and moulds the*
>   *group through communication skills.*

Resistance to change

For resisting change itself, Caruth (1974) puts forward the suggestion that there are a number of key factors. The first is that change, in any form, is perceived as a threat. 'It is seen as a source of frustration, an obstacle which prevents an individual from satisfying a basic need.' It is precisely this fact that is a function of one of the key hypotheses of this research. It is not so much that the need is not satisfied rather it is threatened.

Caruth continues to detail what he considers are more specific reasons, the first being economic security where the change is seen as threatening the source of income. Depersonalisation is a further cause for resistance to change, as Caruth puts it, '...if he feels that the change will carry with it the notion of powerlessness, loss of autonomy, or a loss of identity with the products of one's efforts'. Job status is a further causal factor where change could imply that some of the trappings associated with the present position might be swept away or reduced in status after the proposed change. Change is also seen as introducing levels of uncertainty about competence and the re-skilling or worse, de-skilling, which may result. All this produces a situation where change is seen as disruptive and also possibly a destructive force to the social group within the workplace and hence people tend to resist such change.

The methods used in such resistance can vary widely from the merely outspoken response to the quiet yet positive sabotage of the implementation. Caruth examines a number of possible methods of opposition, ranging from open aggression to the spreading of spurious rumours about the new implementation and the effect it will have on staff. Some will just withdraw and lend no support to the new changes, which in turn leads to the situation where there is no sense of involvement and responsibility shown towards the new system. Others will be totally negative expressing the opinion that the new system will never work and should never have been introduced in the first place. Finally there is the point when the person withdraws totally and this will often lead either to the person being transferred to another department or even leaving the company altogether. Any implementation process that results in the kind of negative outcomes as described by Caruth is already a failure as it has resulted in lose–lose rather than win–win.

This situation is full of potential danger for the change agent as he, or she, has an interest in the success of the implementation. It is important therefore to overcome the barriers to change. Top of the list according to Caruth is that of communicating with the interested parties. If the people who are going to be affected by the change are directly involved with the change from the outset then they have a stake in the development of the system. The people in the organisation are important sources of information, often knowing the underlying methods of operation that exist in the company. It is important that this participation is honest. As Caruth explains

If participation is used simply to placate employees they will very quickly see through the ruse and resistance, perhaps more fiercely than ever, will soon develop.

Hence when such change is being considered the management should consult with employees from the earliest point. Again Caruth states

> *Management should carefully explain the reasons for the change, how it will be implemented, what requirements the new system will impose, the benefits to employees, etc.*

Caruth also emphasises that the negative as well as the positive aspects should be explained. In conclusion he writes

> *People can be conditioned to accept change as a normal occurrence if rewards for acceptance are positive. People will seek out opportunities for change if the right climate has been created by management ... the majority of people will come to accept change if they are allowed to participate in developing the change, if management communicates openly with them concerning change, and if they are taught to accept change as a way of life.*

Leonard-Barton and Kraus (1985) also accept that resistance to change is a major factor and argue that there are two main types of resistance. These they define as firstly overt resistance, which forms as a function of mistakes, or issues that have been overlooked within the implementation plan and secondly tacit resistance, which is a function of underground feelings, which develop into action against the implementation.

The politics of change

The issue of power, the use of power in organisational contexts, and the problems that surround it have been researched and discussed by writers such as Handy (1985), Drucker (1980), Kakabadse and Parker (1984), Pfeffer (1981 and 1992), Lee and Lawrence (1985) and Morgan (1986). What is central to all of these commentators is the fundamental importance of recognising the political dimension associated with organisations and in particular that of change management. The individual who is tasked with managing the change process must recognise the power dimension of what he or she is doing and the likely impact it will have on themselves and others within the organisation.

The person tasked with change has to be aware of whatever power he or she has and be able to determine how this power might be properly used. Leonard-Barton and Kraus (1985) call this person the product champion who will nurture and attempt to anticipate opposition from the person they call the assassin who will equally try to destroy innovation. This the assassin can sometimes do with one careful shot, which means that the champions have to marshal their forces carefully. The most common reasons for this opposition are the fear of de-skilling, loss of power or lack of personal benefit. Leonard-Barton and Kraus feel that a good implementation plan should 'try to identify where a loss of power may occur so that managers can anticipate and possibly avert any problems arising from that loss'. Thus any innovation must offer an obvious advantage over the old system or there will be little incentive to use it. The implementers have considerable power at their fingertips, which can be seen in two primary aspects, positional power and personal power.

Leonard-Barton and Kraus also identify one more character, the hedger. These people sit on the fence waiting for signals that can give them some idea of which way the implementation is going. These people tend to avoid risk and can be found at any level within the organisation. The best way to counter their influence, according to Leonard-Barton and Kraus, is for those in charge of the implementation to send out the right signals so that the hedgers are in no doubt. This can take almost any form from a speech or presentation to a simple quiet word. It is also crucial that the managers at all levels are speaking the same words at the right time. Finally, Leonard-Barton and Kraus consider the important step to be that 'Managers ... bring the criteria used to judge the performance of the users of the innovation into conformance with the demands of the new technology'.

They conclude by saying that the task of converting hedgers is not an easy one to achieve but it is the most inescapable. Indeed

> ... as the competitive effects of new technologies become even more pronounced, the work of implementing those technologies will increasingly pose for managers a distinctive set of challenges – not least the task of creating organisations flexible enough to adjust, adapt and learn continuously.

Power is a key feature of change programmes. Etzioni (1964) discusses the difference between position power and personal power, the distinction springing from his concept of power as the ability to induce or influence behaviour. Etzioni postulates that the best situation for a leader is when he or she has both personal and position power. It is then necessary to consider the use of this power as to whether it will result in success or effectiveness or both. Hersey and Blanchard (1972) consider this by saying

> Success has to do with how the individual or group behaves. On the other hand, effectiveness describes the internal state or predisposition of an individual or group and thus is attitudinal in nature. If an individual is interested only in success, he tends to emphasize his position power and uses close supervision. However, if he is effective he will depend also on personal power and be characterised by more general supervision. Positional power tends to be delegated down from the organisation, while personal power is generated from below through follower acceptance.

This leads Hersey and Blanchard to the conclusion that

> *... a manager could be successful, but ineffective, having only short-run influence over the behaviour of others. On the other hand if a manager is both successful and effective, his influence tends to lead to long-run productivity and organisational development.*

Change/implementation models and conflict resolution examined

Katz and Kahn (1978), like Checkland, developed their approach to organisations with a clear association with open systems theory. A key feature of their work is the careful definition of aspects such as cycles of input, throughput and output in a systems framework. They also recognise the different levels of systems and the interrelationships that exist within the system. It is precisely this relationship that is at the core of the effect–cause–effect logic of the TOC/TP. Through the connection of the logic the TOC/TP attempts to reveal the true causality that exists within the system, and thus the core problem. Katz and Kahn also recognise the importance of conflict and dynamic outcome of such. They cite a number of potential sources of conflict, but offer little in the way of conflict resolution. They do suggest that '... conflicts can have both dysfunctional and functional consequences'.

Katz and Kahn go into the area of conflict in some detail, to them 'conflict requires direct resistance as well as a direct attempt and influence or injury'. Although focusing on conflict at an organisational level they do suggest that there are three commonly used concepts of conflict: conflict of interest, competition and conflict itself by which they imply incompatible interaction. This last concept of conflict is precisely the type of conflict most identified with this research. They go on to argue that 'every aspect of organisational life that creates order and co-ordination of effort must overcome other tendencies to action, and in that fact lies the potentiality for conflict'. With respect to change, they suggest that

> *Organisational change is necessary for survival, but an organisation with no internal resistance to change would be no organisation at all; it would move in any direction, and in response to any suggestion. Change and resistance to change, however, mean conflict.*

It is the recognition that conflict is inevitable that is so encouraging. While many people are trying to avoid conflict, they are actually trying to avoid what is natural in organisations. The key to the reality of conflict lies in the

137

| Decision making | Implementation | Consolidation | Expansion |

Fig. 6.2. Hutchin change model (1986)

knowledge that the conflict can be used to good effect, much in the way that Follett (1995) suggested. This means that some form of conflict resolution is vital to on-going improvement and change.

Therefore the reason why many improvement programmes fail to achieve the expected targets and benefits claimed of them is the absence of a conflict resolution mechanism. These conflicts can occur at almost any point of an improvement process but are most prevalent at two key points. The first is when the core problem of the area under review has been identified and verbalised, and the second during implementation. These two points are part of any improvement process. In 1986, following some research completed at Leicester Polytechnic, I described such a process as comprising decision making–implementation–consolidation– expansion, shown in Fig. 6.2.

The process was assumed to proceed smoothly through each stage and, at the time, the research confirmed the importance of the first stage as being the most crucial to the final outcome. This first stage included the basic decisions about what was required and the decision to go ahead made. Central to the ability to make the right decision were factors such as what must be done? Which areas are the most important? and what level of investment is required? The implementation, which included any training that might be required and the necessary preparation within the organisation for the new systems being installed, followed this.

The third stage, consolidation, came once the initial implementation had been completed. At this point the system was either moving to a successful implementation or significant problems were being experienced. If these problems could not be overcome then the final stage, that of expansion, could not be reached. This stage was concerned with the growth of the new system within the organisation into new areas and perhaps beyond into other aspects of the organisation.

The final stage in this analysis of the literature is to consider what happens when changes have taken place. Every time a change is implemented in an organisation something can be learned. It may be from failure or from success, but either way there is an opportunity for learning to take place. This learning is vital if the organisation is to learn from mistakes, and to ensure that they are not made a second time. What is equally clear is that many organisations miss this opportunity.

Organisational learning

The final section of this chapter covers the area of organisational learning. In the field of managing change, a key feature of successful change programmes is the way in which the organisation learns from what has happened. This may apply to either the individuals concerned on their own or collectively or to the organisation as a whole. Whichever is applicable, and it may be both; the opportunity to learn from the experience should not be missed.

Kolb, Rubin and McIntyre (1971) outlined such a learning model as shown in Fig. 6.3. They observe that

> ... *this learning cycle is continuously recurring in living human beings, Man continuously tests his concepts in experience and modifies them as a result of his observation of the experience. In a very important sense, all learning is re-learning and all education is re-education. Second the direction that learning takes is governed by one's needs and goals. We seek experiences that are related to our goals, interpret them in the light of our goals, and form concepts and test implications of these concepts that are relevant to our felt needs and goals. The implication of this fact is that the process of learning is erratic and inefficient when objectives are not clear.*

Checkland (1981) in his application of soft systems methodology (SSM) also considers the importance of learning. He argues that the methodology of SSM is a learning system in itself. If the process is being used to address

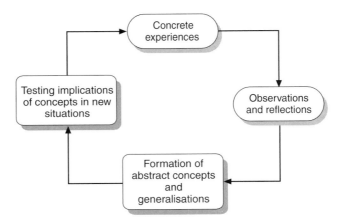

Fig. 6.3. Kolb learning cycle diagram

problems, in particular soft, fuzzy, unstructured problems then as far as Checkland is concerned that

> ... the methodology is a learning system, and in tackling unstructured problems, could only be a learning system, rather than a prescriptive tool, is due to the special nature of human activity systems.

This raises the question of why do people do what they do? Checkland argues that this is a function of the view the individual has of the world. Their actions are conditioned by their thinking. The interpretation of what is happening is also conditioned by this view of the world. Checkland uses the term 'Weltanschauung' to describe this, although he usually abbreviates it to 'W'. Checkland describes this in more detail when he writes

> We attribute meaning to the observed activity by relating it to a larger image we supply from our minds. The observed activity is only meaningful to us, in fact, in terms of a particular image of the world or Weltanschauung, which in general we take for granted.

This suggests that the W any one individual has determine actions. Another term for W is paradigm. For many years a particular paradigm may be appropriate for a particular set of circumstances. However, reality changes and with that change pressure is placed on the ruling paradigm, until it gives way to new thinking, a new paradigm. This is what Kuhn (1970) defines as 'paradigm shift'. Hence, given a set of problems, the way in which the problem is addressed is a function of the W or paradigm of the problem solver. If the solution is successful, the paradigm continues to be valid. If the solution is unsuccessful then the paradigm eventually faces a major challenge to the assumptions of validity that lie behind it.

Checkland observes this when he writes

> It is characteristic of us that we cling tenaciously to the models which make what we observe meaningful. We celebrate Newton and Einstein as the very greatest scientists precisely because they forced the establishment of new Ws. Both were able to establish hypotheses which survived severe tests and hence became public knowledge, and were based on revolutionary frameworks, on Ws different from the prevailing ones of their time. . . . The Ws of an individual man will in fact change through time as a result of his experiences. And the Ws of a group of men perceiving the same thing will also be different. It is because of these two facts that there will be no single description of a 'real' human activity system, only a set of descriptions, which embody different Ws. In a certain sense, human activity systems do not exist; only perceptions of them exist, perceptions which are associated with specific Ws.

For Checkland the power of SSM is in bringing about the ability to break out of one W and move to another.

An analysis of change control models

Within most systems the normal procedure is for some kind of input to the system to trigger, or lead to, an output of the same system. There is also usually some form of feedback to give a degree of control to the system.

The control mechanism usually serves to regulate the input in line with the output in order to prevent the system moving to an out of control position. This model (Fig. 6.4) is a fairly standard description of such systems. It is equally applicable to human activity systems or to most other types of systems. It does have limitations, however, particularly when one important aspect of the system is the ability to learn and thus avoid errors in the future. This ability to learn is a necessary function if the ruling paradigm of the organisation requires changing at any time. This ability to change paradigms has been discussed at great length by Argyris (1990, 1992, and 1993), Argyris and Schon (1996) and also by Argyris, Putnam and Smith (1985). They have developed a number of key concepts within the process of organisational learning.

These include the distinction between espoused theory and theory in action; the nature of single and double loop learning; and the relationship these all have for organisational learning. The model developed by Argyris (1992) can be seen in Fig. 6.5.

Argyris notes

> *Learning is defined as occurring under two conditions. First, learning occurs when an organisation achieves what it intended; that is, there is a match between its design for action and the actuality or outcome. Second, learning occurs when a mismatch between intentions and outcomes is identified and it is corrected; that is a mismatch is turned into a match.*

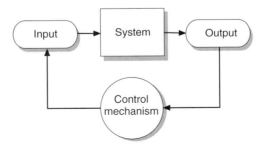

Fig. 6.4. Basic control model

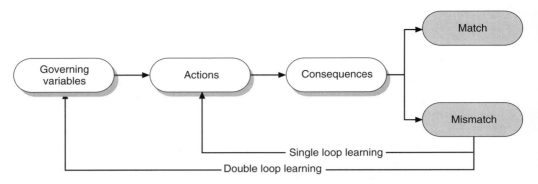

Fig. 6.5. Single/double loop learning (Argyris, 1992)

Referring to the models, Argyris defines them in the following way:

Single loop learning occurs when matches are created, or when mismatches are corrected by changing actions. Double loop learning occurs when mismatches are corrected by first examining and altering the governing variables and then the actions. Governing variables are the preferred states that individuals strive to 'satisfice' when they are acting.

When change is necessary, Argyris suggests that single loop learning is very much in line with the model first introduced by Lewin (1948) comprising three stages, these being unfreezing, changing to a new pattern and then refreezing.

However, the model proved to be inadequate when change of a deeper nature was required. Thus the need for double loop learning is established. Argyris (1992) argues that

Such significant changes require changes in the organizational governing variables and master programs, that is double loop changes. But double loop changes cannot occur without unfreezing the models of organizational structures and processes now in good currency. These models, in turn, cannot be unfrozen without a model of a significantly different organizational state of affairs; otherwise, toward what is the organization to change?

There is, however, an important aspect to what Argyris is saying here. The current model, although still working is no longer valid. This implies that people are using an approach, which they already know is not the one that is required. This leads to the distinction between espoused theory and theory in use. Argyris (1992) argues that

One of the paradoxes of human behaviour, however, is that the master program people actually use is rarely the one they think

they use. Ask people in an interview or questionnaire to articulate the rules they use to govern their actions, and they will give you what I call their 'espoused' theory of action. But observe these same people's behavior, and you will quickly see that this espoused theory has very little to do with how they actually behave. . . . When you observe people's behavior and try to come up with rules that would make sense of it, you discover a very different theory of action – what I call the individual's 'theory in use'. Put simply, people consistently act inconsistently, unaware of the contradiction between their espoused theory and their theory in use, between the way they think they are acting and the way they really act.

At this point Argyris returns to the governing variables when he argues that

. . . most theories in use rest on the same set of governing values. There seems to be a universal human tendency to design one's actions consistently according to four basic values

1. *to remain in unilateral control*
2. *to maximize winning and minimize losing*
3. *to suppress negative feelings*
4. *to be as rational as possible – by which people mean defining clear objectives and evaluating their behavior in terms of whether or not they have achieved them.*

The purpose of all these values is to avoid embarrassment or threat, feeling vulnerable or incompetent. In this respect, the master program that most people use is profoundly defensive.

This last element of Argyris is fundamental. If the change is of some significance then double loop learning is likely. If that is the case then the governing variables must also change. This is the same as the paradigm shift of Kuhn or the change in W of Checkland. However, Argyris suggests that at this point the whole approach of the individual becomes negative. If at the same time they are also aware of the conflict between the espoused theory and the theory in use, the individual can now find himself, or herself, in a degree of difficulty. They have to make decisions about change and whether they are prepared to take that challenge.

The link to the paradigm lock

First introduced earlier in the book (chapter 2), this is a fundamental barrier to change, which cannot be ignored or treated with disdain. The

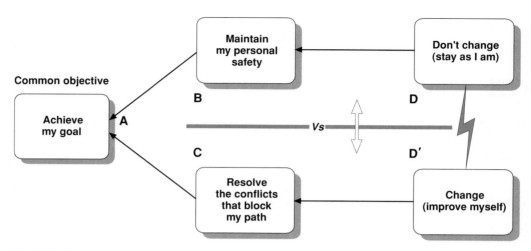

Fig. 6.6. The composite cloud (first attempt)

individual is locked into the dominant paradigm and is clearly going to stick with it. The development of the cloud itself did not take too long, the research process gave me about 50 clouds to work with and it did not take too long to build the basis cloud.

The problem lay in what to place in the **B** box. This proved to be the most difficult as those participating found it very hard to say what was preventing them from implementing the proposed solution. The difficulties surrounding the phrasing of the content of the **B** box was considered to be an appropriate point for such widening of the involvement.

Once the process of determining the entry for each box in the composite cloud was completed, the cloud was then checked by reading the logic in the manner described in chapter 5. The strength of the cloud was also checked through the nature of the cross-connections. The first iteration of the cloud is shown in Fig. 6.6.

The second step is the surfacing of the assumptions that lie beneath each arrow. This in turn checks the logic of the cloud once more and when the statements as written in each box fail to capture the entities in the individual boxes, the composite cloud is rewritten and thus upgraded. In this case the entities did not quite capture the essence of what concerned the individuals. The next diagram, Fig. 6.7, shows the second iteration of the composite cloud.

The cloud reads as follows: 'In order to achieve my goal in life I must break the constraints that block me', and then, 'in order to break the constraints that block me, I must change'. However, 'In order to achieve my goal I must remain in/retain control' and then 'in order to remain in/ retain control I must stay as I am'.

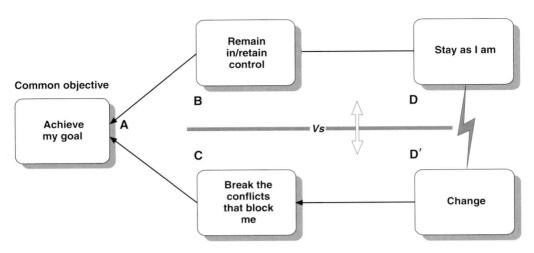

Fig. 6.7. The composite cloud (second attempt)

The logic of the cloud is as follows. The individual knows that there is a goal to be achieved. One necessary condition for the achievement of that goal is the need to break the constraints that block the way forward. In certain cases this will involve personal change. The change may be in what is done, the responsibilities assumed, the actions that are necessary and so on. However, the change is seen by the individual as a threat to the other necessary condition, that written in the box **B** of the cloud. The logic is that if the individual has to change then control is lost. That cannot be countenanced and therefore the change does not take place. The next step of surfacing assumptions was carried out focusing in particular on the arrow **B–D**, as this is the arrow that should be focused on, with the result as shown in Fig. 6.8.

The key issue that arose out of this analysis was that the current paradigms of the individual, what Argyris (1992) calls the 'governing variables' and what Checkland (1981) calls 'W', were key factors in governing the decision-making process of the individual, especially when considering changes which affected them personally. In each of the clouds of chapter 5, the current paradigms of the individuals were effectively locking them into the *status quo* thus preventing the proposed change from taking place.

What the composite cloud was demonstrating was the ability to verbalise this locking of the individual into their current paradigm. This led to the description of the composite cloud as the *paradigm lock cloud*. Once the cloud had been verbalised in this way, it was possible to return to both the data collected and some of the people involved, checking whether this cloud captured their predicament. This led to the final

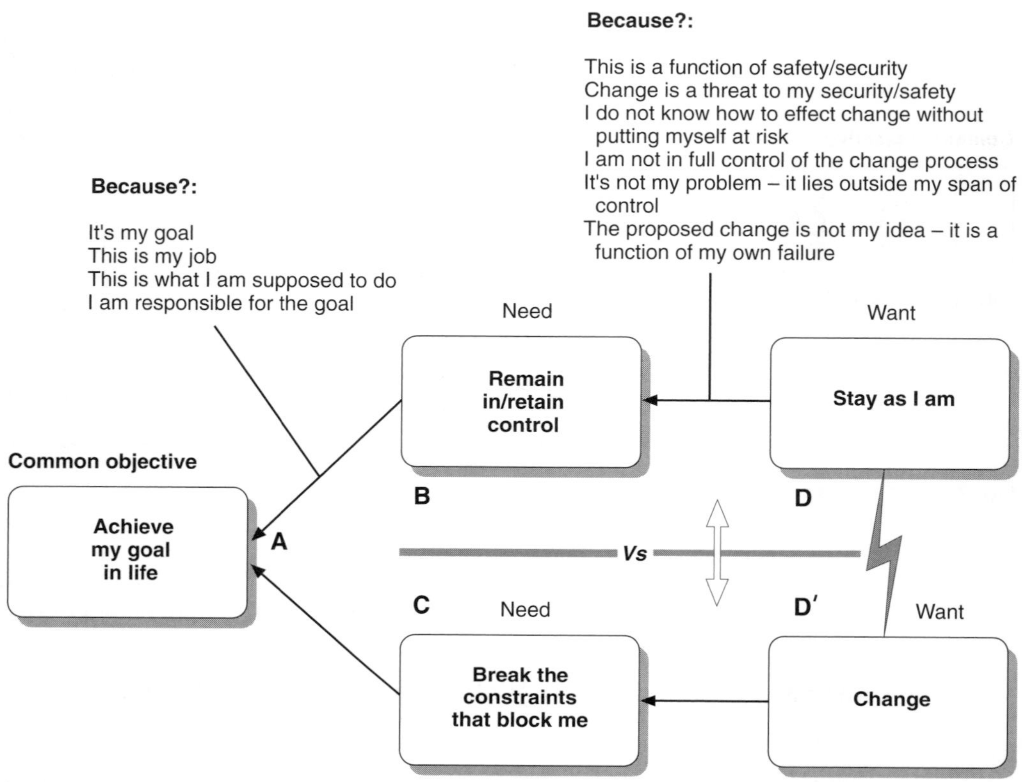

Because?:

This is a function of safety/security
Change is a threat to my security/safety
I do not know how to effect change without
    putting myself at risk
I am not in full control of the change process
It's not my problem – it lies outside my span of
    control
The proposed change is not my idea – it is a
    function of my own failure

Because?:

It's my goal
This is my job
This is what I am supposed to do
I am responsible for the goal

Need

Want

Common objective

Achieve my goal in life	A

Remain in/retain control — B

Stay as I am — D

Vs

C — Need

D'

Want

Break the constraints that block me

Change

Fig. 6.8. Assumptions related to the composite cloud

step, which was to examine whether the apparent dysfunctional, irrational behaviour on the part of the individual, which was giving rise to the inactivity and defensive behaviours could be explained by the paradigm lock cloud (PLC). This analysis led to the final iteration of the cloud as shown in Fig. 6.9.

The personal paradigm is simply defined as the set of values or beliefs that the individual holds about a particular subject, hence 'with respect to' contained within the boxes **D** and **D'**. It might be about cost accounting, it might be about production scheduling, it might be about any area where some change is required. The key element is that the individual holds this set of values and is not minded to change them. For each of the people in chapter 5 there were many areas of their lives where change was not an issue or a blockage. However, when the proposed change was seen by the individual to significantly threaten a personal paradigm then the lock came into play. They often knew that change has to take place. They had, in this research, been involved in both the analysis of the original problem and the development of the solution. However, come

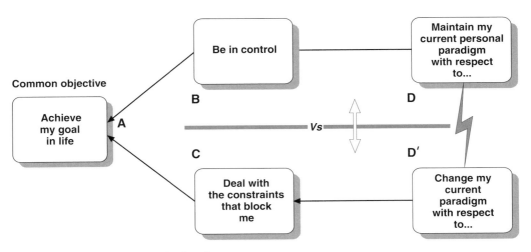

Fig. 6.9. The paradigm lock cloud

the moment of implementation, when no other avenue was open to them, the full power of the lock was activated.

Linking the paradigm lock cloud to models of change

In Fig. 6.2 a simple model of change was described. Given the determination of a clearly verbalised obstacle in the form of the PLC, the next step was to review the model in the light of the cloud. The original model consisted of four stages, which were now seen as insufficient to describe the processes taking place. This led to the first extension of the original model, shown in Fig. 6.10.

Through further extension (Fig. 6.11), the increased level of detail highlighted, with greater clarity, the stages of problem solving. One element was still missing, namely, that in many cases the model did not proceed past stage 5, solution implementation. Indeed often no implementation took place at all. It was this factor that had led to the description of the PLC as one source of non-completion. Noting this omission led to the next enhancement of the change model as shown in Fig. 6.11. This shows the change process surrounding the implementation stage in more detail and in particular the arrow under which the PLC acts.

It should be noted that there were occasions when problems did occur, but they fell into the category of a rational or functional explanation for the non-performance. In those cases the people involved simply returned to the part of the process that required further work and carried it out and then moved to a successful conclusion. Examples of these included times when the original analysis was found to be wanting in some respect,

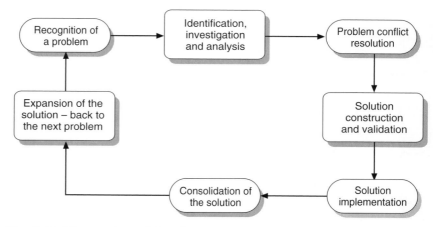

Fig. 6.10. First review of the change model

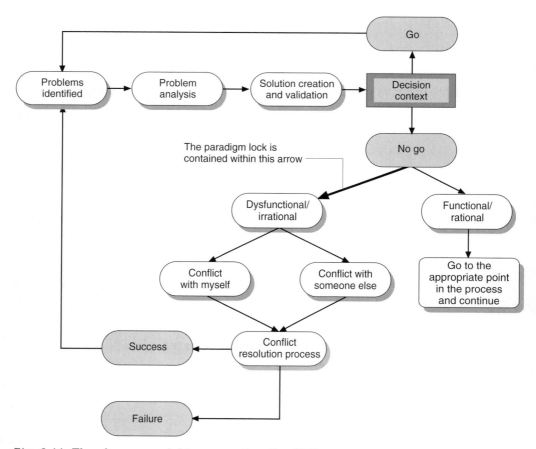

Fig. 6.11. The change model incorporating the PLC

perhaps inadequate analysis of the market, or poor product development. In other cases it was the recognition that the problems being addressed were in fact not the correct or the most urgent ones and this forced those involved to return to an earlier stage on the model and re-evaluate their work and possibly change tack altogether. These reasons for not progressing with the implementation were seen as rational. However this research has shown that there is also an irrational, dysfunctional force operating that prevents the implementations taking place. This led to the revision of the change model to include that factor.

The obstacle to change is the paradigm lock cloud acting at the point shown in a change process.

The impact of the paradigm lock cloud on the individual

People for whom this acts as an obstacle feel that there is no way out of their situation. The logical effects in terms of outcomes of their situation are very clear; often the causal relationships behind the logic are not.

If they are involved in attacking a problem it is reasonable to assume that some form of change will have to take place. Where the change does not affect a personal paradigm the individual has no difficulty in accepting the proposed change, it simply takes place. Where the proposed change violates a personal paradigm then the PLC begins to take effect. Decisions, which can be made in line with the personal paradigm, are relatively simple to make, those, which undermine the personal paradigm become extremely difficult. If it is possible, then the decision to go ahead with the change is postponed. If this is not possible, then other delaying tactics are brought into play. Hence the personal paradigm becomes a barrier to change, which leaves the individual unable to deal with the original problem or constraint. This in turn means that the original goal, written in box **A**, cannot be achieved, because dealing with the problem/constraint is a necessary condition. This leaves only what is written in box **B** as the goal of the individual. Within the case studies many involved found that their primary purpose at work was to protect their personal paradigm.

The PLC as the implementation constraint related to change

This constraint is not that of a project, it is more of an organisational constraint. Within the companies taking part within the research the initial people involved were highly supportive of what the critical chain approach was trying to achieve. Once the implementation process was under way it soon became apparent that other factors not initially recognised started to make their impact felt. This was always associated with the need for support and commitment from other people who had not been involved in the original decision process. The ability to determine whether the PLC was operating or not is difficult to achieve. The first step is to assume that it is not operating and consider the reasons for not moving forward which are more related to the lack of subordination. This is, for no other pretext, a safer place to examine reasons given the high levels of stress when paradigms are being challenged.

The importance of subordination in change management

Subordination is the third of the five steps of focusing contained within the TOC approach. It is also the most difficult to achieve. Time and time again we come across people who find it almost impossible to subordinate to the decisions that have been taken. The conflict of subordination cloud, first introduced in chapter 2, is a profound cloud. It sits at the core of many implementation problems in almost all aspects of change management. This is an area where time spent on surfacing the real assumptions of the individual will pay handsome dividends. There is a great deal of time to be allocated here, and there is often pressure to press ahead and gain the expected benefits of the investment. The reality is that applying that level of pressure is counterproductive and will almost always result in severely reduced benefit, and greater degrees of animosity from the team.

Preparing the organisation for change – the use of measures

It is a truism to say that show me the measures and I'll show you how I'll work, yet in many companies we find that the measurement system dominates behaviour patterns. Therefore if change is necessary it is also necessary to examine the measurement systems to ensure that they align with the proposed changes. If they do not, change the measures. Within all the companies the financial measurement system played a major role in determining how investment was made, how purchasing was carried out. This would often occur even though the reasons for the increased investment were entirely in line with bringing the project back into line with the objectives set for it. In some cases additional expense was refused even though the likely penalty far outweighed the expense being sought.

Overcoming the resistance to change that people have through the use of the TOC/TP tools

This was dealt with in more detail in the previous chapter. Remember for every change being proposed, people will have objections, more importantly they will have a degree of fear about what the proposed changes mean for them personally. The TOC uses the two specific techniques introduced earlier to deal with these problems, the first is called the prerequisite tree (PRT) and the other the transition tree (TRT). The prerequisite tree is a logical structure designed to facilitate the easier implementation of ambitious targets. For example, within the critical chain implementation process a number of key features must be implemented and for each one there will be obstacles. These are listed by the implementation team and checked to ensure that they are actual obstacles to the specific feature being addressed. Once that has been done the next step is to determine what must be implemented in order to overcome the obstacle. This is called the *intermediate objective* and is more usually abbreviated to IO. The IO then serves as part of the building block of the solution. Each IO is examined in turn for logical dependency on other IOs and for time dependency. The result is a time dependent, logical structure for the achievement of the overall feature. Each feature is subjected to this analysis and once every feature has been analysed the overall implementation plan is prepared. This can then be carried out straight away.

Sometimes an IO requires the same level of analysis as the feature itself. This leads to the creation of nested loops within the overall plan. Sometimes the step by step approach is necessary to properly determine which step follows which step, and why. This is the application of the transition tree. Both of these tools are described by Scheinkopf (1999) and Leach (2000).

The five stages of successful change

The model developed as part of the research identified five steps in the process of successful change. The first is gaining consensus on the problem. The second is gaining consensus on the direction of the solution. The third is gaining consensus on the benefits of the solution. The fourth is the ability to overcome all reservations and the fifth is the ability to make it happen. This is a process of on-going improvement which I describe as a wheel of change.

The TOC wheel of change

This wheel (Fig. 6.12) represents the change process contained within the TOC approach. Each part of the wheel contains specific TP tools. It also reflects the fact that it is not necessary to start at the beginning! If a suggestion is put forward, an idea proposed, it is by definition an injection, therefore it must break a cloud – is it possible to construct the cloud,

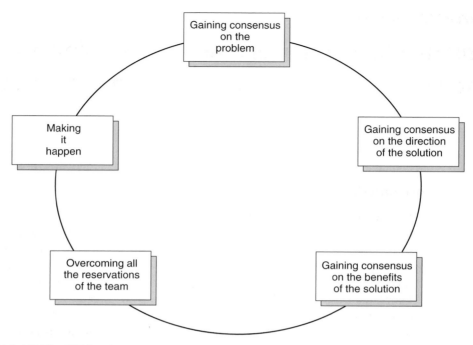

Fig. 6.12. The TOC wheel of change

and surface the assumptions to check that the idea/injection actually breaks the cloud? It is also possible to check further by using the idea and the cloud it breaks as the basis for constructing the CA and possibly the full analysis called the current reality tree. If someone suggests a target, it can seen as a desirable effect (DE), which must have a complementary UDE, and from that the cloud can be checked and so the process goes on. Any starting point can be placed on the TP road map and an analysis developed from there. This of course does not come overnight, and the practitioners of TOC worldwide have accepted the need to practice these tools as often as possible. One clear outcome of the research is that only with mastery of the tools does real, on-going improvement come. What the TOC/TP represents is a coherent, focused, problem-solving tool containing a high degree of rigour and logic. If the intention is to seek specific solutions to specific problems then the application of the TOC/TP seems to be most appropriate.

CHAPTER SEVEN

Linking change management to gaining an enterprise focus in project-based industries

Introduction

Managing change is one key dimension of improving performance, enterprise focus is another, this chapter brings these two key issues together. It would appear that many writers, commentators and researchers have come to the same conclusion, that having an enterprise focus is a necessary condition for success in the markets of today. There are others who have held this view for over 50 years and possibly more. It is simply the choice between the big picture, the global view versus the small picture, the local view. It is an excellent choice cloud as there will be times when it is perfectly reasonable to choose concentration on the local, and at other times choose concentration on the global. One aspect is vital, however, whether the view you are taking today is either one or the other, all decisions must take note of the global view.

Within the enterprise it is the global view that counts the most. Trying to run the organisation only through local management of small discrete entities is never going to be enough. The assumption that if all the little bits are working well then the whole must also be working well has been proven to be erroneous so many times it is beyond belief that it still has currency. There will be times when the focus will be on a small aspect of the business, but always in the context of the whole, this is the basis for the systemic view of the organisation. This book hopefully presents a strong case for the global view being dominant. The lessons learnt from the case studies come to the conclusion that without an enterprise wide view of the organisation the results will always be sub-optimal.

Changing the paradigm – moving to a systemic approach

The problem with the enterprise-focused approach is the number of entrenched ideas that abound in industry today. If there is to be a real shift to such a new focus, then old ideas will have to be challenged, and those found wanting discarded. This is no easy task. Many of the rules of running a successful business have been developed in the old paradigm, indeed they are still taught in many schools, colleges and universities. These rules, and the measurements that support them, enshrine the *status quo* and fail to allow the kind of challenge necessary to break free and change the way people operate inside organisations. The combination of rules and measurements determine behaviour and thus the loop is reinforced. One person who has spoken widely about these issues was Deming. Deming (1994) defined a system as 'a network of interdependent components that work together to try to accomplish the aim of the system'. This is little different from the notion of the revenue chain introduced in chapter 1 with the objective of ensuring that the organisation achieves the goal set for it. Deming goes on to argue that

> *It is management's job to direct the efforts of all components towards the aim of the system. The first step is clarification, everyone in the organisation must understand the aim of the system and how to direct his efforts towards it.*

This is echoed by Joiner (1994) when he argued that 'optimising separate pieces of the chain destroys the effectiveness of the whole'.

Deming goes on to state that

> *An important job of management is to recognise and manage the interdependencies between components. Resolution of conflicts and removal of barriers to co-operation, are responsibilities of management.*

Remember, if we are managing the whole in line with the goal of the system, then it is the whole that must be optimised, the components will not be. When the individual components try to operate optimally, without reference to the rest, Deming argues that the result is harm which stems from the resulting internal competition and conflict, and from the fear that is created in such environments. To Deming

> *the obligation of any component is to contribute its best to the system, not to maximise its own production, profit or sales, nor any other competitive measure. Some components may operate at a loss to themselves in order to optimise the whole system including the components that take a loss.*

This is a fundamental point in enterprise-based organisations, especially those involved in projects. The constraint of the system, the project, the critical chain must operate at its best. Here is where meeting, or beating, the estimated duration pays handsome dividends provided quality has not been sacrificed for speed. The non-critical paths, the feeding paths need not, indeed should not, operate in the same manner, they must subordinate properly to the constraint.

Paradigms and the effect they have

In his book *Paradigms and barriers*, Margolis (1993) sets out to show the impact paradigms can have on thinking and behaviour. He talks about paradigms being changed, or shifted, as changing habits of the mind. We all have habits, physical habits are there for all to see. Margolis argues that there is also a set of habits which are part of our thinking, hence the term habits of the mind. Such habits are not necessarily wrong, or act as a barrier. Margolis writes

> *In the overwhelming majority of cases, habits of the mind serve as facilitators of effective thinking, not as barriers, just as physical habits ordinarily facilitate effective performance, not impede it.*

In the world of projects and project management, the training, the procedures, even the body of knowledge that has been created over the years are all part of the process leading to a habit of mind. Although such habits cannot be observed directly, their consequences can. This can be in terms of how the person responds to questions, new ideas and challenges to normal practice.

The difficulty comes when, as people are normally unaware of such habits, and thus not aware of when the habit is ceasing to be of value, the individual is faced with a requirement to change. Margolis argues

> *So, compared to physical habits, habits of the mind must be hard to detect and identify, which means that clinging to a habit of mind will be complicated by our ordinary lack of conscious sense that a habitual response is in play.*

When two opposing habits of mind meet the result is often disaster. Margolis describes such events when he states

> *Incompatible habits of mind block communication, easily invoke resentment and distaste and frustration, all of which would tend to reinforce a natural propensity toward coordination of habits across individuals and make breaking socially shared habits of mind harder than breaking with socially shared physical habits.*

In the common clash between the local and global perspectives of organisations this can have severe effects. Many organisations with substantial vested interest in maintaining local optima, cost world approach to financial measurements have poured scorn onto the heads of those who argue for a throughput-based, global optima, and vice versa. The nature of the attacks; the power of the rhetoric, have all served only to make matters worse. What is missing is a degree of objectivity, and a pathway between what appears to be two conflicting habits of mind. In some cases it will transpire that one side or the other is wrong. Equally, there will be times when both sides can be accommodated, and what is missing is a bridge. This is what Smith (2000) argues for in her work on financial measurements.

Margolis sums this up beautifully when he writes

> So with respect to habits of mind, situations must arise in which intuitions conflict, and we cannot see easily, or at all why they conflict. In addition neither side sees any reason to try to see things as their adversaries do: rather the puzzle for this lies in why their adversaries cannot see how weak their case is. Hence difficulties arise that have no strong analogue for the case of physical habits. Arguments that seem powerful to one side seem unimportant to the other. What looks like a striking insight to one side looks like perverse illusion to the other.

This last point by Margolis offers a powerful insight to why people have so much difficulty when faced with an almost total requirement for change. If such change clashes with their own habit of mind, developed over many successful years of work, then it is no surprise that they fail to see why they should change. They are locked into the paradigm of today irrespective of any argument for change, no matter how powerful, or true it might be. It is this dimension that surfaced many times during the research phase of this book, and also within many of the implementation of TOC applications.

An analysis of the critical chain approach as a paradigm shift

Critical chain, to be successful, is based on a number of assumptions. The first set is related to the nature of projects and what they are trying to achieve for both the individuals involved and the organisation as a whole. This is as much a question of leadership as anything else. Without clear leadership, usually from the top, although leadership can come from almost any part of an organisation, there is no point in trying to change to a new and challenging paradigm. There has to be a clear need. This also assumes that there is a champion for such change, someone who is prepared to drive the issue through almost regardless of personal cost. The approach to network creation, to the role measurements, to the rules and procedures within the project environment, will have to change if the real benefits are to be realised.

Buffer management will challenge both the notion of keeping people busy, and the assumption that you should rarely allow the budget to be overspent, and if it is then the project is deemed a failure. Budget is probably the worst measurement for projects ever devised, focusing on throughput is far more valuable, far more effective, far more lucrative, but difficult for some people to readily accept.

In purchasing the old models are simply not applicable. The idea of pushing down the supplier, keeping him waiting for payment, treating him only as a necessary evil rather than a full partner, these all have to be ditched in order to really take advantage of what is on offer. Different rules now apply and trying to keep to the old ones only imprisons the organisation in a straightjacket of the mind, and condemns it to possible collapse.

The new reality is one of great opportunity for those who can change the quickest, those lightest on their feet, those more able to challenge and change where necessary. For those who cannot achieve this level of flexibility there is only one real result, you will be taken over, or taken out by those who can.

Organisational constraints and projects

At this point it is clear that the interactive nature of the project and the organisational constraints is the dominant feature, which must be addressed. For projects the need to switch to a critical chain approach is essential, but without the associated change in management methodology the project shift will not work. In combination these two make life very difficult. Yet they must be overcome if we are to achieve a real consensus on the benefits that enterprise focus can bring to the organisation and its people. Joiner (1994) echoes that core of the TOC enterprise approach when he argues that 'by pushing for deeper levels of fix, we are leveraging our efforts, attempting to remedy problems by attacking one deeply rooted cause'. He concludes by adding

> *Every system, no matter how large or how small, has steps, pieces, or components that, when improved, substantially improve the perfor-mance of the system as a whole. These leverage points are small in number but big in effect.*

This is very much in line with what was being discussed in previous chapters concerning the TOC enterprise approach and the rigour used to determine those core drivers, which limit the capability of the system to improve. This notion of the leverage point is a fundamental aspect of this enterprise thinking. It goes straight to the heart of the revenue chain concept, from supply base to the market. Joiner describes this well when he writes

> *Identifying leverage points and thinking about what's best for the system as a whole can help overcome long-standing barriers to improvement. Defining the aim of the system and communicating*

> *that aim to all employees will help people determine their priorities and enhance their customer focus. But the impact of systems thinking can extend even further. By thinking in terms of systems, by drawing larger and larger loops, we can help shape the future. As you and your organisation begins to expand the boundaries of your systems to include your customers, suppliers, communities and competition, think about expanding the time boundary as well.*

This takes the whole dimension to a new level. Joiner is arguing for a coherent, focused and collaborative approach to business management, a view shared by many in the TOC environment.

This book is arguing that such a vision as that of Joiner is no pie in the sky wishful thinking. It is in fact becoming a necessary condition for real long-term growth. The TOC has a pivotal role to play in all of this. It is the process of focusing on the strategic leverage point. To my mind although the TOC as a product has proved to be difficult to sell, TOC as a process is most surely not. Just as Intel scored a huge marketing hit with the slogan 'Intel Inside' for the PC market, similarly for the enterprise solution market the key to success will be the slogan 'TOC Inside'. It may not be visible, but it will be there nonetheless. This is very much linked to the buyer's perception of value.

Joiner reflects on the buyer's perception of value when he writes

> *To deliver world-class quality to our customers, we must understand their perceptions of value. We must help employees to understand the customer's target; we must develop methods for reliably getting closer and closer to that target, reducing the variation about the target.*

Compare this with the discussion described earlier when what the project team were going to deliver was decided with no reference to the client at all, simply on an analysis of what they felt they could deliver. Within the enterprise focus there must be clear visibility of what both the client wants, and what the organisation can deliver.

Enterprise systems are capable of making customer information available throughout the whole organisation. But if the people do not know how to interpret and use the data, do not know the importance of the data, then the focus will be lost. If the current systems preclude such innovation then is it any wonder that many enterprise resource planning systems are failing to deliver. This message lies at the heart of Goldratt's book *Necessary but not sufficient* (2000). One important point that Joiner (1994) makes is

> *A customer-focused approach to strategic and annual planning begins with customers and not financials. It begins with the attitude 'how do we provide even higher value to our customers with lower cost?'.*

The change on methodology that critical chain requires

Critical chain is on the surface a very easy approach to use within project management. At the same time there appears to be a strong defence of the old ways, the old paradigms of managing projects.

Software forms a core element of critical chain implementations. Currently two software companies have products in the market and there are others entering the market in the near future. This book is not the place to analyse software products, although there are observations that can be made, and some suggestions for the future, in particular for enterprise-focused organisations.

Developing the schedules for single and multi-project environments demands a change in the way the networks are created. The use of templates to shorten the network activity, the development on internal specialists, the creation of mentors to support the project teams and the resource managers all play their part. Once the agreement to use critical chain in multi-project environments has been agreed, the first step is to construct the network of activities for each project and then to schedule the strategic resource across all projects. Once the schedules have been created and validated they must be communicated to all the project team. Visibility is a primary element of success here.

The ability to link the performance of all projects to the bottom line is now mandatory. If the revenue chain requires further analysis this is the time to do it. Once under way it is essential that all people making decisions can see the impact of those decisions on the revenue chain. At the same time, the management need to keep an eye on all aspects of the management of the projects in order to check whether new organisational constraints are being created, or surfaced.

CHAPTER EIGHT

Recommendations for success

Introduction

It is always making yourself a hostage to fortune with such a chapter title but this book is for those who really want to make a difference, really make it happen. In the light of that assumption having some recommendations seems not unreasonable. The starting point for any project-based organisation seeking to really move forward is the state of the current revenue chain and the location and type of constraint currently operating. If the normal business and financial performance indicators are in play, such as market share and shareholder value, profit and return on investment, then there is usually plenty of scope for substantial improvement. Seagate used the TOC approach to dominate the 15 000 rpm disc drive market.

The role of technology in addressing a limitation

When technology is being considered as part of the solution, there is one clear assumption that can be made; the technology itself must diminish a limitation within the existing system. This limitation must also have existed before the technology is implemented. This usually means that over time, the existence of the limitation has led to a number of rules and/or procedures being developed and used on a regular basis. The TOC approach to project management, critical chain, and the associated management methodology addresses a limitation within the world of project management. The area of project management has for many years struggled with uncertainty in meeting project end dates, keeping to budget, retaining customer features and so on. This book has highlighted many of those problems through the case studies. Critical chain is the technology that was developed, with associated software, to address these issues.

Equally, given the nature of the implementation, and the enterprise nature of both the approach and the methodology, a limitation in the context of the overall performance of the organisation is also addressed and this represents an organisation wide change.

The lesson from this research, and the work of people such as Goldratt in his book entitled *Necessary but not sufficient* (2000), is that whenever a new technology is implemented without reference to the existing systems of management, both formal and informal, no real change will take place in terms of bottom-line improvement. In other words the limitation is not properly addressed and its impact not lessened. Within critical chain many of the features of the solution have been implemented in many organisations but not all are having the kind of success they should. There are many success stories, of course, but others are still not achieving the level of bottom-line performance they should. This research has shown that in most cases the cause of this is the inability of the team within the organisation to properly address the systemic problems.

Enterprise resource planning (ERP) offers an opportunity to have before any one manager, or group of managers, far more information than ever before and in real time. Information that can be used to see what is happening within the organisation at the time is now readily available. Information that can be used to make better decisions than before, coupled with the ability to carry out risk assessment prior to implementing solutions. However the experience of the last two or three years is that ERP systems have yet to make substantial impact on bottom-line performance.

Ptak and Schragenheim (2000) make the point that ERP allows for information about all aspects of the organisation to be made available, and from that to examine the impact at any point in the revenue chain. The limitation in the use of all this data is the view the individual has of the organisation. For years the only view has been that of the immediate vicinity of the individual. They have only been able to see what they have direct control over and perhaps a little more in adjacent areas or functions. ERP changes that, an enterprise focus changes all that, now it is possible to view the whole organisation.

But at the same time many of the measures; many of the procedures and many of the rules have not changed. In the small, local perspective the rule has always been to maximise your area and expect everyone else to do the same. The assumption has been that the sum of all the individual maxi-misations will equal the total performance of them all. This is a local optimisation focus. With this focus came a whole series of rules and procedures, all to ensure that the focus is maintained.

Now along comes a new technology, an enterprise technology, encap-sulated in both approaches such as critical chain and in management methodologies such as the TOC, which challenge this local focus. The ability to see the whole chain through the information systems, to make decisions at a local level but where the global impact can be readily ascertained is now here. The technology delivers the global focus; the local focus is no longer required. But the systems of the local focus have not been taken away. The rules and measures of the local focus still dominate. Therefore the benefits of the new technology are lost in the cries of anguish about the loss of local focus. The ruling, dominant paradigm takes over, the habits of many years rule, the change cannot be countenanced, and the opportunity lost.

This is not a unique feature of ERP or critical chain. Almost every week the press announces yet another plant closure, more people being made redundant, more layoffs and, for those involved, more misery. Nearly always these closures, this ability to treat people as something to be discarded with little or no apparent emotion is treated as part of the deal, a necessary condition for the survival of the rest. It is nearly always nonsense. It is simply that those at the top, those making the decision have no other model to work with, no other process to assess the situation and develop a different response. They are as much trapped as those being laid off. What this book is arguing for is another way forward. A fundamental shift in thinking, from local to global, from the individual to the team.

This is, of course, an excellent example of certainly a conflict of subordination, where the senior team are changing to the new focus but others at all levels within the organisation fail to follow. It is also the area of the paradigm lock, where once more people at all levels simply cannot accept the changes that are necessary for the organisation to develop and grow. They now hold the organisation back.

Making it happen

The final step in the TOC wheel of change is called *making it happen*. I chose this because when I completed the research into the paradigm lock, I felt that the most important part of any change programme, assuming all the previous aspects contained within the wheel had been done properly, was still the need to make it happen, to actually implement the solution through to completion.

It starts with top management support and commitment. There is nothing new here. Writers such as De Bono and Handy have been arguing for this for many years. All this research has shown is that the importance of this support has not diminished indeed has grown. There is a need for real strategical thinking in the boardrooms of many organisations, not just those with a project bias. Within civil engineering there has been for far too long an air of 'It'll be alright on the night'! Too little has been challenged, and too many people have settled for just maintaining the *status quo*. Many initiatives have foundered on the fact that the remit given has been too limited, usually to avoid offending the many senior managers and board members within the industry.

Investors do not come out of the research with a great deal of credit. Their level of understanding of the industry is often restricted to just money, and always in the short term. They have to protect their investment, but as long as the payback periods are kept short, and the financial indicators continue to use the cost-based models of yesterday they remain a poor source of real ideas and innovation.

The decision-making process needs to be examined. All too often the political dimension of organisations dominates the process of decision making. Vested interest takes over the process and emotions soon run high.

Measurements need to be aligned throughout the organisation. The ability to determine just what is happening at any one moment of time is vital, and that means that the measurement process aids people to know what decisions to make, not to hinder them.

Re-engineering the organisation became synonymous with cutting people. It became a process of reducing the organisation to a rump, getting rid of those who failed to come up to expectation and often crippling the company at the same time. However, with a TOC focus it is possible to properly re-engineer the organisation. What the TOC approach is seeking is a unique lever point. That area within the organisation where real leverage can be obtained and substantial improvement in performance achieved. In many implementations around the world people are reporting excellent bottom-line upgrade in a short period of time. General Motors (GM) at one of the TOC world conferences, held in St Paul, MN, stole the show by claiming a $500 million improvement on the bottom line, in this case additional sales inside of one month. How many other companies can claim that level of improvement in that space of time? Why did this happen? The presenters simply said that all they did was to properly implement the TOC approach – this time in production, and the results just followed. They searched for that strategic lever point, found it, and used it.

Of course to do this means the company has to have a robust thinking process. It has to have a capability to look in the mirror and challenge all rules, all measures, all procedures and to be ready and able to throw out the ones that do not work and replace them with ones that do.

Transferring techniques from wherever they have worked is the future. In sport it is common for coaches to examine what is happening in other sports and where applicable bring them across to their sport. In manufacturing the transfer of ideas from Japan into the western manufacturing arena has proved to be one of the greatest challenges, and successful transfers of recent years. Earlier I described the example of GM who, at a conference in July 2000, noted that they had all the key ideas and techniques, they had Kaizen, they had lean, they had total quality management (TQM) and so on, and the thread that bound them all together, the process that made it all happen was the rigorous use of the TOC and the TP. This is not an isolated example, GM are not alone in searching around the world for ideas about how to maintain their position.

To improve across the whole enterprise demands clear thinking. To improve across the whole enterprise requires people who are prepared to make it happen. To achieve this level of success demands people who can challenge their own beliefs and if necessary throw away all that has brought them to the positions they hold now, and switch to a better way. This is not difficult, not if the goal is worth going for.

Recognising the impact of the constraint, perhaps even designing the constraint into the organisation, is vital for long-term growth. Trying to buck the constraint is like trying to buck the markets, you cannot do it for long if at all.

The importance of having the goal clearly understood, and the necessary conditions for the achievement of the goal also clearly understood, is vital if people are to pull together. So many companies seem to spend their time going round and round in circles. Without a clear vision, without a clear map of where the organisation is going, is it any wonder that people make it up for themselves, and never seem to worry if the various goals and objectives conflict. In the absence of real leadership why should they be concerned?

When considering the paradigm lock I proposed three major changes to the way the individual functioned if they were to overcome the power of the lock. The first was 'taking responsibility and being accountable for the results of my actions'. This meant quite simply that whatever happened as a result of one of my decisions, I had to be accountable, I had to stand up and take the responsibility. I should not try to shift the blame elsewhere. If I am in charge of that part of the organisation, or indeed the whole organisation, then it is simply down to me and no-one else. The second element to successfully overcome the lock was to subordinate to both the goal of the organisation and the constraint that limits its ability to deliver. This assumed that the goal had been properly defined and communicated. It also assumed that I had the ability to know what the constraint was. If I knew neither then it was my responsibility to find out what they were and work accordingly. The final element was to develop the ability to both give and respond to leadership. Leadership is not the same as management. Leadership can come from anywhere, from high level, from low level, it might even come from outside the organisation. The question is how sensitive am I to leadership? Do I recognise it when I should? Leadership is not always found in the boardroom, neither is it missing from the shop floor. If I can implement these three elements then the impact of paradigm lock is substantially reduced if not broken altogether.

Final thoughts

This book has described what is happening in a number of project-based industries. The case studies have been derived from civil engineering, high-tech product development and a number of manufacturing companies using project management techniques to effect process improvement. Many had success. Many knew they could have done so much better. The reasons for failing to maximise improvement have been discussed in detail. The sense of frustration people had about the continual politicking, the never-ending and almost always fruitless meetings were a constant theme. Many people felt anger when months of valuable work was thrown away as senior management once more rearranged the deck chairs on what was now a sinking ship. Too often senior people relied on the size of the company and the strength it had in the market-place to make it through the bad times, the occasions when non-performance was clearly visible.

Yet for all these problems and frustrations many kept going. Continuing to overcome obstacle after obstacle, resolve conflict after conflict in order to achieve the goal. To them the goal of the company and their own personal goal was the same. To never be satisfied with current performance, to achieve the dream – however it might be defined – to leave behind a legacy that others could build on and take pride in. To generate and disseminate knowledge, knowledge about systems, about organisations, about people, and to use that knowledge to create wealth is what they wanted to leave behind. It is to them that this book is really dedicated.

References

Argyris, C. (1990). Inappropriate defenses against the monitoring of organization development practices. *J. Applied Behavioral Science*, **26** (3) 299–312.

Argyris, C. (1992). *On organizational learning.* Blackwell, Cambridge, MA.

Argyris, C. (1993). *Knowledge for action.* Jossey-Bass, San Francisco.

Argyris, C. and Schon, D. A. (1996). *Organizational learning II.* Addison-Wesley, Wokingham.

Argyris, C., Putnam, R. and McLain Smith, D. (1985). *Action science.* Jossey-Bass, San Francisco.

Atkinson, P. E. (1985). Who should manage change? *Management Services*, **29** (2), 14–15.

Bach, R. (1997). *Illusions.* Dell Fiction, New York.

Brooks, E. (1980). *Organizational change. The managerial dilemma.* Macmillan, London.

Buchanan, D. A. and Huczynski, A. A. (1985). *Organisational behaviour.* Prentice Hall International.

Burns, T. and Stalker, G. (1966). *The management of innovation.* Tavistock, London.

Caruth, D. L. (1974). Basic psychology, *J. Systems Management*, Feb., 10–13.

Checkland, P. (1981). *Systems thinking, systems practice.* John Wiley, Chichester.

Checkland, P. and Scholes, J. (1991). *Soft systems methodology in action.* John Wiley, Chichester.

Corbett (1998). *Throughput accounting.* North River Press, Great Barrington, MA.

Cox and Spencer (1998). *The constraints management handbook.* St Lucie Press, Boca Raton, FL.

Deming, W. E. (1986). *Out of the crisis.* MIT CAES, Cambridge, MA.

Deming, W. E. (1994). *The new economics.* CAES, Cambridge, MA.

Dettmer, H. W. (1997). *Goldratt's theory of constraints.* ASQC Quality Press, Milwaukee, WI.

Drucker, P. F. (1980). *Managing in turbulent times.* Butterworth Heinemann, London.

Emery, F. E. and Trist, E. L. (1965). The causal texture of organizational environments. *Human Relations*, **18**, 21–32.

Etzioni, A. (1964). *Modern Organizations.* Prentice-Hall, Englewood Cliffs, NJ.

Feigenbaum, A. V. (1991). *Total Quality Control.* 3rd edn., McGraw Hill, New York.

Follett, M. P. (1995). *Prophet of management.* Harvard Business School Press, Classic, MA.

Gilmore, M. and Smith, D. J. (1996). Set-up reduction in pharmaceutical manufacturing: an action research study. *Int. J. Operations and Production Management,* **16** (3), 4–17.

Goldratt, E. M. (1990). *Theory of constraints.* North River Press, New York.

Goldratt, E. M. (1997). *Critical chain.* North River Press, Great Barrington, MA.

Goldratt, E. M. (2000). *Necessary but not sufficient.* North River Press, Great Barrington, MA.

Goldratt, E. M. and Cox, J. (1984). *The goal.* Gower, London.

Goldratt, E. M. and Fox, R. (1986). *The race.* North River Press, London.

Handy, C. B. (1985). *Understanding organisations.* Penguin, London.

Hersey, P. and Blanchard, K. H. (1972). The management of change. *Training and Development Journal,* Jan., 6–10.

Hersey, P. and Blanchard, K. (1988). *Management of organizational behavior,* Prentice-Hall International, Englewood Cliffs, NJ.

Hutchin, C. E. (1986). Paving the way for technological change, *Proc. 3rd Int. Conf. Human Factors in Manufacturing,* IFS 35–42.

Imai, M. (1986). *Kaizen.* Random House, New York.

Ishikawa, K. (1990). *Introduction to quality control.* Chapman and Hall.

Joiner (1994). *Fourth generation management.* McGraw-Hill, New York.

Kakabadse, A. and Parker, C. (eds) (1984). *Power politics and organisations.* Wiley, Chichester.

Katz, D. and Kahn, R. L. (1978). *The social psychology of organisations.* 2nd edn., John Wiley, USA.

Katzenbach, J. R. (1995). *Real change leaders.* Random House.

Kolb, D. A., Rubin, I. M. and McIntyre, J.M. (1971). *Organizational psychology. An experiential approach.* 2nd edn., Prentice-Hall, New Jersey.

Kuhn, T. S. (1970). *The structure of scientific revolutions.* 2nd edn., University of Chicago Press.

Lawrence, P. R. and Lorsch, J. W. (1967). *Organisation and environment.* Harvard University Press, Cambridge, MA

Leach, L. P. (2000). *Critical chain project management.* Artech House, Norwood, MA.

Leonard-Barton, D. and Kraus, W. A. (1985). Implementing new technology. *Harvard Business Review,* Nov.–Dec., 102–109.

Lee, R. and Lawrence, P. (1985). *Organisational behaviour – politics at work.* Hutchinson, London.

Lewin, K. (1948). *Resolving social conflicts.* Harper and Row, New York.

Likert, R. (1967). *The human organisation.* McGraw Hill, New York.

Margolis, H. (1993). *Paradigms and barriers.* University of Chicago Press, Chicago.

McMullen Jr., T. B. (1998). *Theory of constraints management system.* St Lucie Press, Boca Raton, FL.

Moore, M. and Gergen, P. (1985). Risk taking and organizational change. *Training and Development J.,* **39** (6), 72–76.

Morgan, G. (1986). *Images of organization.* Sage Publications, London.

Newbold, R. C. (1998). *Project management in the fast lane.* St Lucie Press, Boca Raton, FL.

Newell, A. and Simon, H. A. (1972). *Human problem solving.* Prentice-Hall, Englewood Cliffs, NJ.

Newman, V. (1995). *Problem solving for results.* Gower, London.

Noreen, E., Smith, D. and Mackey, J. T. (1995). *The theory of constraints and its implications for management accounting.* North River Press, Great Barrington, MA.

Pascale, R. (1991). *Managing on the edge.* Penguin. London.

Pettigrew, A. and Whipp, R. (1996). *Managing change for competitive success.* Blackwell Business, Oxford.

Pfeffer, J. (1981). *Power in organizations.* Ballinger Publishing, Cambridge Mass.

Pfeffer, J. (1992). *Managing with power.* Harvard Business School Press, Boston, MA.

Porter, M. E. (1980). *Competitive strategy.* Free Press, New York.

Ptag, C. and Schragenheim, E. (2000). *ERP tools, techniques and applications for integrating the supply chain.* St Lucie Press, Boca Raton, FL.

Ruelle, D. (1991). *Chance and chaos.* Penguin, London.

Schonberger, R. J. (1982). *Japanese manufacturing techniques.* Free Press, New York.

Schonberger R. J. (1986). *World class manufacturing.* Free Press, New York.

Scheinkopf, L. (1999). *Thinking for a change.* St Lucie Press, Boca Raton, FL.

Skinner, W. (1974). The focused factory. *Harvard Business Review*, May–June, 113–121.

Smith, D. (2000). *The measurement nightmare.* St Lucie Press, Boca Raton, FL.

Srikanth, M. L. and Robertson, S. A. (1995). *Measurements for effective decision making.* Spectrum Publishing, Guilford, CT.

Swain, M. and Bell, J. (1999). *The theory of constraints and throughput accounting.* Irwin McGraw Hill, USA.

Tichy, N. M. (1983). Managing organisational transformations. *Human Resource Management*, **22** (1/2).

VanGundy A. B. (1988). *Techniques of structured problem solving.* 2nd edn., Van Nostrand Reinhold, New York.

Womack, J. P. Jones, D. T. and Roos, D. (1990). *The machine that changed the world.* Rawson Associates, New York.

Wooldridge, E. (1982). Negotiating technological change. *Personnel Management*, Oct., 40–43.

Index